Lecture Notes in Computer Science 12087

More information about this series at http://www.springer.com/series/7410

Marco Baldi · Edoardo Persichetti ·
Paolo Santini (Eds.)

Code-Based Cryptography

8th International Workshop, CBCrypto 2020
Zagreb, Croatia, May 9–10, 2020
Revised Selected Papers

Springer

Editors
Marco Baldi
Department of Information Engineering
Marche Polytechnic University
Ancona, Italy

Edoardo Persichetti
Department of Mathematical Sciences
Florida Atlantic University
Boca Raton, FL, USA

Paolo Santini
Department of Information Engineering
Marche Polytechnic University
Ancona, Italy

ISSN 0302-9743 ISSN 1611-3349 (electronic)
Lecture Notes in Computer Science
ISBN 978-3-030-54073-9 ISBN 978-3-030-54074-6 (eBook)
https://doi.org/10.1007/978-3-030-54074-6

LNCS Sublibrary: SL4 – Security and Cryptology

This Springer imprint is published by the registered company Springer Nature Switzerland AG
The registered company address is: Gewerbestrasse 11, 6330 Cham, Switzerland

Preface

Fostered by the ongoing NIST's standardization process of post-quantum cryptographic primitives, the research area of post-quantum cryptography is experiencing a dramatic speed-up in these years. Cryptographic primitives relying on the hardness of decoding a random-looking error-correcting code are notoriously resistant to quantum computer-based attacks, and indeed the branch of code-based cryptography today represents one of the most promising lines of research in the area of post-quantum cryptography.

The Code-Based Cryptography Workshop (CBC) was born in 2009 as an informal forum following the initiative of a group of researchers mainly based in Europe. Since then, it has become an important event for a growing academic community in the area of code-based cryptography. The 6th edition (CBC 2018), hosted by the Florida Atlantic University in 2018, was the first one organized in the USA. The 7th edition (CBC 2019) was held in Darmstadt, Germany, in co-location with Eurocrypt 2019, and represented another turning point for the conference with the introduction of submitting works for publication in the form of a post-proceedings volume. The goal of this first edition of CBCrypto was to build on the success of the previous editions of CBC and continue the hybrid publication model which was utilized therein. The choice to move to a new nomenclature was made to accommodate the ever growing interest and differentiate it from the previous series of informal events.

CBCrypto 2020 was supposed to take place in Zagreb, Croatia, during May 9–10, 2020, co-located with Eurocrypt 2020. Due to the COVID-19 global pandemic, the workshop was moved to an online-only format, and was held through live streaming on May 9, 2020. Nevertheless, it proved to be a great success, with over 200 registered participants attending the talks from around the world. The program was enriched with two invited presentations by the internationally recognized researchers Shay Gueron and Alessandro Barenghi. Furthermore, the program included two contributed talks presenting recent research and works in progress.

This book collects the 7 contributions out of 10 submissions that were selected for publication by the Program Committee through careful peer review. These contributions span all aspects of code-based cryptography, from design to implementation, including studies of security, new systems, and improved decoding algorithms. As such, the works presented in this book provide a synthetic yet significant overview of the state of the art of code-based cryptography, laying out the groundwork for future developments.

We wish to thank the Program Committee members and the external reviewers for their hard and timely work. We are also very grateful to our sponsor, Oak Ridge

Associated Universities (ORAU), for their generous support. Finally, we thank the Organizing Committee of Eurocrypt 2020 for the inclusion of CBCrypto 2020 among Eurocrypt 2020 co-located events.

May 2020

Marco Baldi
Edoardo Persichetti
Paolo Santini

Organization

Organizing Committee

Marco Baldi	Università Politecnica delle Marche, Italy
Edoardo Persichetti	Florida Atlantic University, USA
Paolo Santini	Università Politecnica delle Marche, Italy

Program Committee

Marco Baldi	Università Politecnica delle Marche, Italy
Gustavo Banegas	Technische Universiteit Eindhoven, The Netherlands
Alessandro Barenghi	Politecnico di Milano, Italy
Emanuele Bellini	Technology Innovation Institute, UAE
Sergey Bezzateev	Saint Petersburg University of Aerospace Instrumentation, Russia
Olivier Blazy	Université de Limoges, France
Pierre-Louis Cayrel	Laboratoire Hubert Curien, France
Franco Chiaraluce	Università Politecnica delle Marche, Italy
Tung Chou	Osaka University, Japan
Alain Couvreur	LIX, École Polytechnique, France
Jean-Christophe Deneuville	École Nationale de l'Aviation Civile, France
Taraneh Eghlidos	Sharif University of Technology, Iran
Shay Gueron	University of Haifa, Israel
Cheikh T. Gueye	University of Dakar, Senegal
Grigory Kabatiansky	Skoltech, Russia
Gianluigi Liva	German Aerospace Center (DLR), Germany
Chiara Marcolla	Technology Innovation Institute, UAE
Gretchen Matthews	Virginia Tech, USA
Giacomo Micheli	University of South Florida, USA
Kirill Morozov	University of North Texas, USA
Gerardo Pelosi	Politecnico di Milano, Italy
Edoardo Persichetti	Florida Atlantic University, USA
Simona Samardjiska	Radboud University, The Netherlands
Paolo Santini	Università Politecnica delle Marche, Italy
John Sheekey	University College Dublin, Ireland
Jean-Pierre Tillich	Projet Secret, Inria, France
Antonia Wachter-Zeh	Technical University of Munich, Germany
Øyvind Ytrehus	University of Bergen, Norway

Additional Reviewers

Junaid Ahmad Khan
Lina Mortajine
Nicolas Sendrier
Violetta Weger

Contents

On the Security of NTS-KEM in the Quantum Random Oracle Model

Varun Maram(✉)

Department of Computer Science, ETH Zurich, Zurich, Switzerland
vmaram@inf.ethz.ch

Abstract. NTS-KEM is one of the 17 post-quantum public-key encryption (PKE) and key establishment schemes remaining in contention for standardization by NIST. It is a code-based cryptosystem that starts with a combination of the (weakly secure) McEliece and Niederreiter PKE schemes and applies a variant of the Fujisaki-Okamoto (Journal of Cryptology 2013) or Dent (IMACC 2003) transforms to build an IND-CCA secure key encapsulation mechanism (KEM) in the classical random oracle model (ROM). Such generic KEM transformations were also proven to be secure in the quantum ROM (QROM) by Hofheinz et al. (TCC 2017), Jiang et al. (Crypto 2018) and Saito et al. (Eurocrypt 2018). However, the NTS-KEM specification has some peculiarities which means that these security proofs do not directly apply to it.

This paper identifies a subtle issue in the IND-CCA security proof of NTS-KEM in the classical ROM, as detailed in its initial NIST second round submission, and proposes some slight modifications to its specification which not only fixes this issue but also makes it IND-CCA secure in the QROM. We use the techniques of Jiang et al. (Crypto 2018) and Saito et al. (Eurocrypt 2018) to establish our IND-CCA security reduction for the modified version of NTS-KEM, achieving a loss in tightness of degree 2; a quadratic loss of this type is believed to be generally unavoidable for reductions in the QROM (Jiang et al. ePrint 2019/494). Following our results, the NTS-KEM team has accepted our proposed changes by including them in an update to their second round submission to the NIST process.

Keywords: Code-based · KEM · Quantum random oracle model · IND-CCA security · NIST standardization

1 Introduction

NIST's post-quantum cryptography (PQC) standardization project reached its second phase when, on 30th January 2019, a shortlist of 26 second-round candidate algorithms was announced – out of which 17 are public-key encryption (PKE) and key establishment schemes, and the rest are digital signature schemes [NIS19]. In a public-key setting, a key encapsulation mechanism (KEM) is considered to be a versatile cryptographic primitive, as it can be used for efficient black-box constructions of secure PKE (via the KEM-DEM paradigm [CS03]), key

M. Baldi et al. (Eds.): CBCrypto 2020, LNCS 12087, pp. 1–19, 2020.
https://doi.org/10.1007/978-3-030-54074-6_1

exchange and authenticated key exchange schemes [BCNP08,FOPS01]. Hence, a majority of these 17 second-round submissions are proposals for KEMs.

Indistinguishability against chosen-ciphertext attacks (IND-CCA) is widely accepted as the standard security notion for KEMs and PKE schemes, but it is usually more difficult to prove than weaker notions of security such as indistinguishability (IND-CPA) and one-wayness (OW-CPA) against chosen-plaintext attacks. Therefore, most of the NIST KEM submissions employ some generic transformations, as studied by Dent [Den03] and Hofheinz et al. [HHK17], to construct an IND-CCA secure KEM from a weakly (OW-CPA or IND-CPA) secure PKE. To be specific, these generic constructions are usually variants of the Fujisaki-Okamoto transformation [FO13], e.g., $\mathrm{FO}^{\perp}$, $\mathrm{FO}^{\not\perp}$, $\mathrm{FO}^{\perp}_{m}$ and $\mathrm{FO}^{\not\perp}_{m}$ (as named in [HHK17]). Figure 1 contains a description of the $\mathrm{FO}^{\not\perp}_{m}$ transformation that, given hash functions $G(.)$ and $H(.)$, turns an OW-CPA secure PKE $(\mathsf{KGen}_{\mathrm{PKE}}, \mathsf{Enc}, \mathsf{Dec})$ to an IND-CCA secure KEM $(\mathsf{KGen}_{\mathrm{KEM}}, \mathsf{Encap}, \mathsf{Decap})$ – see Subsect. 2.3 for definitions of PKEs and KEMs. (Also, $\mathsf{Enc}(\mathsf{pk}, \mathbf{m}; G(\mathbf{m}))$ denotes that $G(\mathbf{m})$ is used as random coins in the encryption of message $\mathbf{m}$ sampled from the message space $\mathcal{M}$.)

$\mathsf{KGen}_{\mathrm{KEM}}$	$\mathsf{Encap}(\mathsf{pk})$	$\mathsf{Decap}(\mathbf{c}, \mathsf{sk}')$
$(\mathsf{pk}, \mathsf{sk}) \leftarrow \mathsf{KGen}_{\mathrm{PKE}}$	$\mathbf{m} \leftarrow\!\$\ \mathcal{M}$	$\hat{\mathbf{m}} = \mathsf{Dec}(\mathsf{sk}, \mathbf{c})$
$\mathbf{z} \leftarrow\!\$\ \mathcal{M}$	$\mathbf{c} = \mathsf{Enc}(\mathsf{pk}, \mathbf{m}; G(\mathbf{m}))$	**if** $\mathsf{Enc}(\mathsf{pk}, \hat{\mathbf{m}}; G(\hat{\mathbf{m}})) = \mathbf{c}$
$\mathsf{sk}' = (\mathsf{sk}, \mathbf{z})$	$\mathbf{K} = H(\mathbf{m})$	**return** $H(\hat{\mathbf{m}})$
return $(\mathsf{pk}, \mathsf{sk}')$	**return** $(\mathbf{K}, \mathbf{c})$	**else return** $H(\mathbf{z} \mid \mathbf{c})$

Fig. 1. IND-CCA secure KEM = $\mathrm{FO}^{\not\perp}_{m}[\mathrm{PKE}, G, H]$.

The other three variants (namely $\mathrm{FO}^{\perp}$, $\mathrm{FO}^{\not\perp}$ and $\mathrm{FO}^{\perp}_{m}$) have slight differences: the subscript m (without m, resp.) means that, in the corresponding transformation, the encapsulated key $\mathbf{K}$ is equal to $H(\mathbf{m})$ ($\mathbf{K} = H(\mathbf{m} \mid \mathbf{c})$, resp.), and the superscript $\perp$ ($\not\perp$, resp.) means explicit[1] (implicit, resp.) rejection of invalid ciphertexts during decapsulation.

Typically, the security of such schemes is analyzed (heuristically) in the *random oracle model* (ROM), introduced in [BR93], where a hash function is idealized as a publicly accessible random oracle. But as pointed out by Boneh et al. [BDF+11], in a post-quantum setting, an adversary could evaluate a hash function on an arbitrary superposition of inputs. This is not captured in the ROM as an adversary is only given a *classical* access to the random oracle. In order to fully assess the post-quantum security of cryptosystems, the *quantum random oracle model* (QROM) was advocated in [BDF+11]. Here, the adversary

[1] In explicit rejection, the symbol "$\perp$" is returned (instead of a pseudorandom key $H(\mathbf{z} \mid \mathbf{c})$ as is the case in implicit rejection) for the decapsulation of invalid ciphertexts.

is allowed to make quantum queries to the random oracle. The above generic KEM transformations ($FO^{\perp}$, $FO^{\not\perp}$, $FO^{\perp}_m$, $FO^{\not\perp}_m$) were initially only analyzed in the ROM by Hofheinz et al. [HHK17] but then later were proven to be secure in the QROM by Jiang et al. [JZC+18] and Saito et al. [SXY18], giving confidence in the NIST KEM candidates that rely on these transformations.

NTS-KEM is a KEM proposal that is shortlisted by NIST for PQC standardization. It is also one of a handful of second-round candidates that are code-based, as it is based on the well-known McEliece cryptosystem [McE78]. NTS-KEM employs a transformation similar to the Fujisaki-Okamoto [FO13] (or Dent [Den03]) transforms to achieve IND-CCA security of its KEM in the ROM. In particular, the transformation looks similar to $FO^{\not\perp}_m$ since, as will be detailed in Sect. 3, NTS-KEM does an implicit rejection of invalid ciphertexts during decapsulation, and when computing the encapsulated keys, the ciphertext is not included in the input to the hash function. But at the same time, NTS-KEM contains significant variations from $FO^{\not\perp}_m$ in its specification, meaning that straightforward application of known QROM security proofs for FO-transformations ([JZC+18,SXY18]) do not work. One of these major variations from $FO^{\not\perp}_m$ is that, during the encapsulation of keys, the message $\mathbf{m}$ to be encrypted is not sampled uniformly from the message space, but is determined from the *randomness* $\mathbf{e}$ that is used in the McEliece-type encryption function.

1.1 Our Contributions

In this paper, we make two contributions:

- We identify a flaw in the IND-CCA security proof for NTS-KEM in the ROM, as described in its initial NIST second round submission. We also propose some changes to the specification of NTS-KEM which, in addition to fixing the flaw, preserve the tightness of the intended ROM proof.
- We present a proof of IND-CCA security for the modified version of NTS-KEM in the QROM. On a high level, our proof is structurally similar to [JZC+18]'s QROM security proof of $FO^{\not\perp}_m$. At the same time, our proof needs to account for significant differences between $FO^{\not\perp}_m$ and the new NTS-KEM specification.

To be specific, we recommend a re-encryption step in the NTS-KEM decapsulation routine (similar in spirit to the FO-type transformations) to account for invalid ciphertexts that may not be implicitly rejected. This change not only fixes NTS-KEM's tight IND-CCA security proof in the classical ROM but also leads to an IND-CCA security reduction in the QROM, only incurring a quadratic loss w.r.t. degree of tightness. This loss might be impossible to avoid [JZM19].

In order to formulate a security proof in the QROM for the modified NTS-KEM, we consider the $FO^{\not\perp}_m$ transformation which is proven to be secure in the QROM. So to devise a proof based on the $FO^{\not\perp}_m$ framework, we view the random bits of the encryption function used in NTS-KEM, i.e., the error vectors $\mathbf{e}$, as "messages" encrypted by OW-CPA secure PKEs in the context of

$\mathrm{FO}_m^{\not\perp}$. Namely, we start with the scheme NTS$^-$, a variant of NTS-KEM as will be defined in Sect. 4, that involves a "randomized encryption" of error vectors during key encapsulation. By doing this, we found it necessary to work with a non-standard security notion called *error one-wayness* or EOW security, as introduced in [ACP+19a], which is an analogue to OW-CPA security but for schemes that mainly process error vectors. Then the challenge is to account for notable differences between the modified NTS-KEM scheme and $\mathrm{FO}_m^{\not\perp}$ in the proof, which includes the fact that to derive the encapsulated keys $\mathbf{K}$ in NTS-KEM, the message $\mathbf{m}$ itself is not hashed but a modified version of it is.

It is worth mentioning that the NTS-KEM team has adopted our proposed changes. On 3rd December 2019, an updated specification of NTS-KEM [ACP+19b] was posted on the website https://nts-kem.io/. Hence, in the rest of this paper, "NTS-KEM" will be used to refer to the updated version of the second round submission to NIST's PQC standardization process, unless stated otherwise.

2 Preliminaries

2.1 Notation

In this section, we outline notation borrowed from [ACP+19a] regarding NTS-KEM. We denote by $\mathbb{F}_2$ the field with two elements, and by $\mathbb{F}_{2^m}$ an extension field of $\mathbb{F}_2$ with 2^m elements. If $\mathbb{F}$ is a field, then $\mathbb{F}[x]$ is the ring of univariate polynomials with coefficients in $\mathbb{F}$. We denote by $\mathbb{F}_2^n$ the n-dimensional vector space with entries in $\mathbb{F}_2$, and by $\mathbb{F}_2^{k\times n}$ the kn-dimensional vector space of matrices with k rows and n columns with entries in $\mathbb{F}_2$. We denote vectors of $\mathbb{F}_2^n$ in bold lowercase, for example $\mathbf{e} = (e_0, e_1, \ldots, e_{n-1}) \in \mathbb{F}_2^n$; and matrices of $\mathbb{F}_2^{k\times n}$ in bold uppercase, for example $\mathbf{G} \in \mathbb{F}_2^{k\times n}$. The *Hamming weight* of a vector $\mathbf{e}$ is the number of non-zero components in the vector and is denoted by $\mathrm{hw}(\mathbf{e})$. Given a vector $\mathbf{e}$ of length n over a field $\mathbb{F}$, and positive integers $\ell < k < n$, we adopt the following notation to denote the partition of $\mathbf{e}$ into three sub-vectors: $\mathbf{e} = (\mathbf{e}_a \mid \mathbf{e}_b \mid \mathbf{e}_c)$, where $\mathbf{e}_a \in \mathbb{F}^{k-\ell}$, $\mathbf{e}_b \in \mathbb{F}^{\ell}$ and $\mathbf{e}_c \in \mathbb{F}^{n-k}$. More generally, if $\mathbf{v} \in \mathbb{F}^{n_1}$ and $\mathbf{w} \in \mathbb{F}^{n_2}$ are vectors over $\mathbb{F}$, we will denote by $(\mathbf{v} \mid \mathbf{w})$ the vector in $\mathbb{F}^{n_1+n_2}$ constructed as the concatenation of $\mathbf{v}$ and $\mathbf{w}$. A *permutation vector* $\mathbf{p} = (p_0, p_1, \ldots, p_{n-1})$ is a permutation of the n elements $\{0, 1, \ldots, n-1\}$. Then given the sequence $\mathbf{b} = (b_0, b_1, \ldots, b_{n-1})$, we denote the permuted sequence $\mathbf{b}' = \mathbf{b} \cdot \mathbf{P} = \pi_{\mathbf{p}}(\mathbf{b})$ such that $b'_i = b_{p_i}$, and the inverse permutation is given by $\mathbf{b} = \mathbf{b}' \cdot \mathbf{P}^{-1} = \pi_{\mathbf{p}}^{-1}(\mathbf{b}')$ such that $b_{p_i} = b'_i$. We denote the length of a vector $\mathbf{x}$ by $|\mathbf{x}|$.

The security parameter is denoted by λ. Given a set X, we denote by $x \leftarrow_{\$} X$ the operation of sampling an element $x \in X$ uniformly at random, and we denote the sampling according to some arbitrary distribution D by $x \leftarrow D$. We denote probabilistic computation of an algorithm A on input x by $y \leftarrow_{\$} A(x)$. $A^{H(.)}$ implies that the algorithm has access to the oracle $H(.)$.

2.2 Quantum Random Oracle Model

We introduce some lemmas in the QROM that will be used to derive the main results of this paper.

Lemma 1 (Simulating a QRO, [Zha12, Theorem 6.1]**).** *Let $H(.)$ be an oracle drawn from the set of 2q-wise independent functions uniformly at random. Then the advantage any quantum algorithm making at most q quantum queries to $H(.)$ has in distinguishing $H(.)$ from a truly random oracle is identically 0.*

Lemma 2 ([SY17, Lemma C.1]). *Let $g_z : \{0,1\}^\ell \to \{0,1\}$ denotes a function defined as $g_z(z) = 1$ and $g_z(z') = 0$ for all $z \neq z'$, and $g_\perp : \{0,1\}^\ell \to \{0,1\}$ denotes a function that returns 0 for all inputs. Then for any unbounded adversary $\mathcal{C}$ that issues at most q quantum queries to its oracle, we have*

$$|\Pr[\mathcal{C}^{g_z(.)}(.) \to 1 \mid z \leftarrow_{\$} \{0,1\}^\ell] - \Pr[\mathcal{C}^{g_\perp(.)}(.) \to 1]| \leq q \cdot 2^{-\frac{\ell+1}{2}}$$

Lemma 3 (Generalized OW2H[2] lemma, [JZC+18, **Lemma 3 condensed**]**).**[3] *Let oracles $\mathcal{O}_1(.), \mathcal{O}_2(.)$, input parameter inp and x be sampled from a joint distribution D, where $x \in \{0,1\}^n$ (the domain of $\mathcal{O}_1(.)$). Consider a quantum oracle algorithm $\mathcal{U}^{\mathcal{O}_1,\mathcal{O}_2}$ which makes at most q_1 queries to $\mathcal{O}_1(.)$ and q_2 queries to $\mathcal{O}_2(.)$. Denote $\ddot{\mathcal{O}}_1(.)$ to be a reprogrammed oracle such that $\ddot{\mathcal{O}}_1(x) = y$, for a uniformly random y in $\{0,1\}^\ell$, and $\ddot{\mathcal{O}}_1(.) = \mathcal{O}_1(.)$ everywhere else. Let $\mathcal{V}^{\ddot{\mathcal{O}}_1,\mathcal{O}_2}$ be an oracle algorithm that on input $(inp, x, \mathcal{O}_1(x))$ does the following: picks $i \leftarrow_{\$} \{1,\ldots,q_1\}$, runs $\mathcal{U}^{\ddot{\mathcal{O}}_1,\mathcal{O}_2}(inp, x, \mathcal{O}_1(x))$ until the i-th query to $\mathcal{O}_1(.)$, measures the query in the computational basis and outputs the measurement outcome (when $\mathcal{U}$ makes less than i queries, $\mathcal{V}$ outputs $\perp \notin \{0,1\}^n$). Define the events E_1, E_2 and probability $P_{\mathcal{V}}$ as follows,*

$$\Pr[E_1] = \Pr[b' = 1 : (\mathcal{O}_1, \mathcal{O}_2, inp, x) \leftarrow D, y \leftarrow_{\$} \{0,1\}^\ell, b' \leftarrow \mathcal{U}^{\mathcal{O}_1,\mathcal{O}_2}(inp, x, \mathcal{O}_1(x))]$$

$$\Pr[E_2] = \Pr[b' = 1 : (\mathcal{O}_1, \mathcal{O}_2, inp, x) \leftarrow D, y \leftarrow_{\$} \{0,1\}^\ell, b' \leftarrow \mathcal{U}^{\ddot{\mathcal{O}}_1,\mathcal{O}_2}(inp, x, \mathcal{O}_1(x))]$$

$$P_{\mathcal{V}} = \Pr[x' = x : (\mathcal{O}_1, \mathcal{O}_2, inp, x) \leftarrow D, y \leftarrow_{\$} \{0,1\}^\ell, x' \leftarrow \mathcal{V}^{\ddot{\mathcal{O}}_1,\mathcal{O}_2}(inp, x, \mathcal{O}_1(x))]$$

Then $|\Pr[E_1] - \Pr[E_2]| \leq 2q_1\sqrt{P_{\mathcal{V}}}$.

---

[2] The one-way to hiding (OW2H) lemma, introduced in [Unr14], provides a generic reduction from a hiding-style property (indistinguishability) to a one-wayness-style property (unpredictability) in the QROM.

[3] We are referring to the latest version of [JZC+18] on the Cryptology ePrint Archive – Report 2017/1096, Version 20190703 – which differs from the conference version in that Lemma 3 no longer requires $\mathcal{O}_1(x)$ to be independent from $\mathcal{O}_2(.)$. Also we would be working with a condensed version of the lemma where we do not need$\mathcal{O}_1(x)$ to be uniformly distributed for any fixed $\mathcal{O}_1(x')$ $(x' \neq x)$, $\mathcal{O}_2(.)$, *inp* and x.

2.3 Cryptographic Primitives

Definition 1. *A Public Key Encryption scheme (PKE) consists of the following triple of polynomial-time algorithms* (KGen, Enc, Dec).

- *The Key Generation algorithm* KGen *takes as input a security parameter* 1^λ *and outputs a public/private key-pair* (pk, sk).
- *The Encryption algorithm* Enc *takes as input a public key* pk *and a valid message* $\mathbf{m}$, *and outputs a ciphertext* $\mathbf{c}$.
- *The Decryption algorithm* Dec *takes as input a ciphertext* $\mathbf{c}$ *and a private key* sk, *and outputs a message* $\mathbf{m}$ *(or an error message indicating a decryption failure).*

For example, McEliece [McE78] proposed a PKE scheme which is based on linear error-correcting codes. Let $\mathcal{C} = [n, k, d]_2$ be such a code over $\mathbb{F}_2$ of length n and dimension k, with minimal distance d. The code $\mathcal{C}$ is capable of correcting at most $\tau = \lfloor \frac{d-1}{2} \rfloor$ errors, and can be described by a generator matrix $\mathbf{G} \in \mathbb{F}_2^{k \times n}$. Then a vector $\mathbf{w} \in \mathbb{F}_2^k$ can be encoded as a codeword in $\mathcal{C}$ as $\mathbf{c} = \mathbf{w} \cdot \mathbf{G} \in \mathbb{F}_2^n$. Now the McEliece scheme could be described as follows.

KGen: Generate a *special* type of $[n, k, d]_2$ linear error-correcting code $\mathcal{C}_\mathcal{G}$[4] with generator matrix $\mathbf{G}' \in \mathbb{F}_2^{k \times n}$ and that is capable of correcting up to τ errors; this special code is defined by a polynomial $G(z) \in \mathbb{F}_{2^m}[z]$ of degree τ. Let $\mathbf{S}$ be a non-singular matrix in $\mathbb{F}_2^{k \times k}$ and $\mathbf{P}$ be a permutation matrix in $\mathbb{F}_2^{n \times n}$, both generated at random. Then, define $\mathbf{G} = \mathbf{S} \cdot \mathbf{G}' \cdot \mathbf{P}$. The public key is given by $\mathsf{pk} = (\mathbf{G}, \tau)$ and the private key is $\mathsf{sk} = (G(z), \mathbf{S}^{-1}, \mathbf{P}^{-1})$.

Enc: To encrypt a message $\mathbf{m} \in \mathbb{F}_2^k$, sample $\mathbf{e} \in \mathbb{F}_2^n$ with Hamming weight τ and output the ciphertext $\mathbf{c} = \mathbf{m} \cdot \mathbf{G} + \mathbf{e} \in \mathbb{F}_2^n$.

Dec: To recover the message $\mathbf{m}$, compute $\mathbf{c}' = \mathbf{c} \cdot \mathbf{P}^{-1} = \mathbf{m} \cdot \mathbf{S} \cdot \mathbf{G}' + \mathbf{e} \cdot \mathbf{P}^{-1}$, and decode $\mathbf{c}'$ using a decoder for $\mathcal{C}_\mathcal{G}$ to recover the permuted $\mathbf{e}$, and hence $\mathbf{m}' = (\mathbf{m} \cdot \mathbf{S}) \in \mathbb{F}_2^k$. Finally, recover $\mathbf{m} = \mathbf{m}' \cdot \mathbf{S}^{-1}$ and output $\mathbf{m}$.

The McEliece PKE scheme achieves a certain notion of security known as *one-wayness* (OW), relying on hardness of the well-known problem of decoding random linear codes. Roughly speaking, the notion states that an adversary cannot recover the underlying message $\mathbf{m}$ from a given ciphertext $\mathbf{c}$. The OW security notion is formalized in Fig. 2, where we write $\{0,1\}^{\mathsf{poly}(\lambda)}$ for the message space, indicating that it consists of bit-strings of some length that depends on some polynomial function of the security parameter.

An adversary $\mathcal{A}$ is said to be a (t, ε)-adversary against OW security of a PKE scheme if that adversary causes the $\mathrm{OW}^{\mathcal{A}}_{\mathsf{Enc}}$ game to output "1" with probability at least ε (where $0 < \varepsilon \leq 1$) and runs in time at most t. A PKE scheme is said to be (t, ε)-secure with respect to a given security notion, such as OW security, if no (t, ε)-adversary exists for that notion.

Definition 2. *A Key Encapsulation Mechanism (KEM) consists of the following triple of polynomial-time algorithms* (KGen, Encap, Decap).

[4] Known as a binary *Goppa* code, as described in [ACP+19a].

$\mathrm{OW}_{\mathsf{Enc}}^{\mathcal{A}}$

$(\mathsf{pk}, \mathsf{sk}) \leftarrow \mathsf{KGen}(1^{\lambda})$

$\mathbf{m} \leftarrow_{\$} \{0,1\}^{\mathsf{poly}(\lambda)}$

$\mathbf{c} \leftarrow \mathsf{Enc}(\mathsf{pk}, \mathbf{m})$

$\mathbf{m}' \leftarrow \mathcal{A}(1^{\lambda}, \mathsf{pk}, \mathbf{c})$

$\mathbf{return}\ (\mathbf{m}' = \mathbf{m})$

Fig. 2. OW-security game for PKE.

- *The Key Generation algorithm* KGen *takes as input a security parameter* 1^{λ} *and outputs a public/private key-pair* (pk, sk).
- *The Encapsulation algorithm* Encap *takes as input a public key* pk *and outputs an encapsulated key and ciphertext* $(\mathbf{K}, \mathbf{c})$.
- *The Decapsulation algorithm* Decap *takes as input a ciphertext* $\mathbf{c}$ *and a private key* sk, *and outputs a key* $\mathbf{K}$ *encapsulated in* $\mathbf{c}$ *(or an error message "*$\perp$*" indicating a decapsulation failure).*

Compared to OW security, the desired notion for a KEM or a PKE scheme is IND-CCA security, i.e. *indistinguishability under chosen ciphertext attacks.* In the KEM version of this security notion, informally, an adversary should not be able to decide whether a given pair $(\mathbf{K}, \mathbf{c}^*)$ is such that $\mathbf{c}^*$ encapsulates $\mathbf{K}$ or if $\mathbf{K}$ is a random key independent of $\mathbf{c}^*$. In addition, the adversary is also given access to a decapsulation oracle that returns the output of $\mathsf{Decap}(\mathbf{c}', \mathsf{sk})$ for any $\mathbf{c}' \neq \mathbf{c}^*$ (and where we assume the adversary never queries $\mathbf{c}^*$ to this oracle, to prevent trivial wins). We denote this capability of accessing an oracle by the adversary as $\mathcal{A}^{\mathsf{Decap}(\cdot,\mathsf{sk})}(1^{\lambda}, \mathsf{pk}, \mathbf{K}_b, \mathbf{c}^*)$ in Fig. 3. Here we write $\{0,1\}^{\mathsf{poly}(\lambda)}$ for the key space, indicating that it consists of bit-strings of some length that depends on a polynomial function of the security parameter λ.

$\text{IND-CCA}_{\mathsf{KEM}}^{\mathcal{A}}$

$b \leftarrow_{\$} \{0,1\}$

$(\mathsf{pk}, \mathsf{sk}) \leftarrow \mathsf{KGen}(1^{\lambda})$

$(\mathbf{K}_0, \mathbf{c}^*) \leftarrow \mathsf{Encap}(\mathsf{pk})$

$\mathbf{K}_1 \leftarrow_{\$} \{0,1\}^{\mathsf{poly}(\lambda)}$

$b' \leftarrow \mathcal{A}^{\mathsf{Decap}(\cdot,\mathsf{sk})}(1^{\lambda}, \mathsf{pk}, \mathbf{K}_b, \mathbf{c}^*)$

$\mathbf{return}\ (b' = b)$

Fig. 3. IND-CCA-security game for KEM.

Formally, a (t, ε)-adversary against the IND-CCA security of a KEM causes the above game to return "1" with probability at least $1/2 + \varepsilon$ (where $0 < \varepsilon \leq 1/2$) and runs in time at most t. We say that a KEM is (t, ε)-secure in the IND-CCA sense if no (t, ε)-adversary exists; NTS-KEM is IND-CCA secure in the classical random oracle model with a tight relationship to the OW-security of the McEliece PKE.

Finally, a KEM (respectively, PKE scheme) is said to be *perfectly correct* if for any public/private key pair $(\mathsf{pk}, \mathsf{sk})$ generated by KGen, we have $\Pr[\mathsf{Decap}(\mathbf{c}, \mathsf{sk}) = \mathbf{K} \mid (\mathbf{c}, \mathbf{K}) \leftarrow \mathsf{Encap}(\mathsf{pk})] = 1$ (respectively, $\Pr[\mathsf{Dec}(\mathbf{c}, \mathsf{sk}) = \mathbf{m} \mid \mathbf{c} \leftarrow \mathsf{Enc}(\mathsf{pk}, \mathbf{m})] = 1$ for any valid message $\mathbf{m}$). For example, the McEliece scheme is a perfectly correct PKE and NTS-KEM is a perfectly correct KEM as shown in [ACP+19a].

3 NTS-KEM Specification

NTS-KEM is a key encapsulation mechanism that can be seen as a mixture of the McEliece and Niederreiter PKE schemes [McE78, Nie86] combined with a transform similar to the Fujisaki-Okamoto [FO13] or Dent [Den03] transforms to achieve (tight) IND-CCA security in the classical ROM. We provide a higher-level overview of the scheme's three main operations – namely *Key Generation*, *Encapsulation* and *Decapsulation* – that is relevant to the main results of this paper (refer to [ACP+19a, ACP+19b] for a more detailed description). Most importantly, the description below also includes our proposed changes to the decapsulation routine.

In the following, (n, τ, ℓ) are public parameters where $n = 2^m$ denotes the length of codewords, τ denotes the number of errors that can be corrected by the code (see McEliece PKE scheme in Subsect. 2.3) and ℓ denotes the length of the random key to be encapsulated. Also k is a value which is chosen such that $k = n - \tau m$ with $\ell < k < n$. NTS-KEM uses a pseudorandom bit generator $H_\ell(.)$ to produce ℓ-bit binary strings; the current version uses the SHA3-256 hash function [NIS15] to implement $H_\ell(.)$.

Key Generation: Without going into details on how the keys are generated, it is sufficient to know that an NTS-KEM public key is given by $\mathsf{pk} = (\mathbf{Q}, \tau, \ell)$ where $\mathbf{Q} \in \mathbb{F}_2^{k \times (n-k)}$ is a matrix used in the encryption of messages during encapsulation, and private key is defined as $\mathsf{sk} = (\mathbf{a}^*, \mathbf{h}^*, \mathbf{p}, \mathbf{z}, \mathsf{pk})$ where $\mathbf{a}^*, \mathbf{h}^* \in \mathbb{F}_{2^m}^{n-k+\ell}$ are used in the decoding algorithm used for decapsulation, $\mathbf{p} \in \mathbb{F}_{2^m}^n$ is a permutation vector and $\mathbf{z} \in \mathbb{F}_2^\ell$ is used in the decapsulation of invalid ciphertexts.

Encapsulation: Given an NTS-KEM public key $\text{pk} = (\mathbf{Q}, \tau, \ell)$, the encapsulation algorithm produces two vectors over $\mathbb{F}_2$ – a random vector $\mathbf{K}$, where $|\mathbf{K}| = \ell$, and the ciphertext $\mathbf{c}^*$ encapsulating $\mathbf{K}$. It uses the following function that acts on n-bit error vectors, and denoted as $\text{Encode}(\mathsf{pk}, \mathbf{e})$, which proceeds as follows.

1. Partition $\mathbf{e}$ as $\mathbf{e} = (\mathbf{e}_a \mid \mathbf{e}_b \mid \mathbf{e}_c)$, where $\mathbf{e}_a \in \mathbb{F}_2^{k-\ell}$, $\mathbf{e}_b \in \mathbb{F}_2^\ell$ and $\mathbf{e}_c \in \mathbb{F}_2^{n-k}$.
2. Compute $\mathbf{k}_e = H_\ell(\mathbf{e}) \in \mathbb{F}_2^\ell$ and construct message vector $\mathbf{m} = (\mathbf{e}_a \mid \mathbf{k}_e) \in \mathbb{F}_2^k$.

3. Perform systematic encoding of $\mathbf{m}$ with $\mathbf{Q}$:

$$\begin{aligned}\mathbf{c} &= (\mathbf{m} \mid \mathbf{m}\cdot\mathbf{Q}) + \mathbf{e}\\ &= (\mathbf{e}_a \mid \mathbf{k}_e \mid (\mathbf{e}_a \mid \mathbf{k}_e)\cdot\mathbf{Q}) + (\mathbf{e}_a \mid \mathbf{e}_b \mid \mathbf{e}_c)\\ &= (\mathbf{0}_a \mid \mathbf{c}_b \mid \mathbf{c}_c)\,,\end{aligned}$$

where $\mathbf{c}_b = \mathbf{k}_e + \mathbf{e}_b$ and $\mathbf{c}_c = (\mathbf{e}_a \mid \mathbf{k}_e)\cdot\mathbf{Q} + \mathbf{e}_c$. Then remove the first $k-\ell$ coordinates (all zero) from $\mathbf{c}$ to output $\mathbf{c}^* = (\mathbf{c}_b \mid \mathbf{c}_c) \in \mathbb{F}_2^{n-k+\ell}$.

NTS-KEM encapsulation is then defined as:

1. Generate uniformly at random an error vector $\mathbf{e} \in \mathbb{F}_2^n$ with $\mathrm{hw}(\mathbf{e}) = \tau$.
2. Compute $\mathbf{k}_e = H_\ell(\mathbf{e}) \in \mathbb{F}_2^\ell$.
3. Output the pair $(\mathbf{K}, \mathbf{c}^*)$ where $\mathbf{K} = H_\ell(\mathbf{k}_e \mid \mathbf{e})$ and $\mathbf{c}^* = \mathrm{Encode}(\mathsf{pk}, \mathbf{e})$.

Decapsulation: [5]The decapsulation of an NTS-KEM ciphertext $\mathbf{c}^* = (\mathbf{c}_b \mid \mathbf{c}_c)$ proceeds as follows.

1. Consider the vector $\mathbf{c} = (\mathbf{0}_a \mid \mathbf{c}_b \mid \mathbf{c}_c) \in \mathbb{F}_2^n$, and apply a decoding algorithm—using the secret parameters $(\mathbf{a}^*, \mathbf{h}^*)$—to recover a permuted error pattern $\mathbf{e}'$.
2. Compute the error vector $\mathbf{e} = \pi_{\mathbf{p}}(\mathbf{e}')$, partition $\mathbf{e} = (\mathbf{e}_a \mid \mathbf{e}_b \mid \mathbf{e}_c)$, where $\mathbf{e}_a \in \mathbb{F}_2^{k-\ell}$, $\mathbf{e}_b \in \mathbb{F}_2^\ell$ and $\mathbf{e}_c \in \mathbb{F}_2^{n-k}$, and compute $\mathbf{k}_e = \mathbf{c}_b - \mathbf{e}_b$.
3. Compute $\mathbf{c}' = \mathrm{Encode}(\mathsf{pk}, \mathbf{e})$. Verify that $\mathbf{c}' = \mathbf{c}^*$ and $\mathrm{hw}(\mathbf{e}) = \tau$. If yes, return $\mathbf{K} = H_\ell(\mathbf{k}_e \mid \mathbf{e}) \in \mathbb{F}_2^\ell$; otherwise return $H_\ell(\mathbf{z} \mid \mathbf{1}_a \mid \mathbf{c}_b \mid \mathbf{c}_c)$.

3.1 Changes to the Initial NTS-KEM Decapsulation [ACP+19a]

The NIST second round submission for NTS-KEM [ACP+19a] does not perform the re-encoding check in the decapsulation algorithm. Specifically, the evaluation of $\mathrm{Encode}(\mathsf{pk}, \mathbf{e})$ in step 3 of the *Decapsulation* operation above is not performed, and instead it only verifies if $\mathrm{hw}(\mathbf{e}) = \tau$ and $H_\ell(\mathbf{e}) = \mathbf{k}_e$ to identify valid ciphertexts. But this may allow some invalid ciphertexts $\mathbf{c}$ to evade implicit rejection by the decapsulation oracle, leading to a possible attack in the IND-CCA security game of NTS-KEM.

To be specific, the initial IND-CCA security proof for NTS-KEM in the ROM [ACP+19a] failed to account for ciphertexts $\mathbf{c}$ which, when given as input to the decoding algorithm used in NTS-KEM decapsulation, result in an error vector $\mathbf{e}$ such that $\mathrm{hw}(\mathbf{e}) = \tau$ and $H_\ell(\mathbf{e}) = \mathbf{c}_b - \mathbf{e}_b$, but $\mathrm{Encode}(\mathsf{pk}, \mathbf{e}) \neq \mathbf{c}$. Because of the correctness of NTS-KEM (as shown in [ACP+19a]), it is not hard to see that such a ciphertext $\mathbf{c}$ is not the result of any valid NTS-KEM encapsulation.[6] This might lead to a potential attack in the IND-CCA security game of NTS-KEM. Given a challenge ciphertext $\mathbf{c}^* = (\mathbf{c}_b^* \mid \mathbf{c}_c^*)$ (along with

[5] Our suggested routine, as adopted in the updated version of NTS-KEM [ACP+19b].

[6] On the contrary, if there exists an error vector $\mathbf{e}'$ with $\mathrm{hw}(\mathbf{e}') = \tau$ such that $\mathrm{Encode}(\mathsf{pk}, \mathbf{e}') = \mathbf{c}$, then because of NTS-KEM correctness, the decoding algorithm should recover error vector $\mathbf{e}'(\neq \mathbf{e})$ when given $\mathbf{c}$ as input.

a key $\mathbf{K}_b$, see Subsect. 2.3 for definitions of security games w.r.t. KEMs), the adversary could possibly construct the above invalid ciphertext $\mathbf{c} = (\mathbf{c}_b^* \mid \mathbf{c}_c)$ by modifying the last $(n-k)$ bits of $\mathbf{c}^*$, such that the decoding algorithm in NTS-KEM decapsulation would recover the error vector $\mathbf{e}^*$ used in the NTS-KEM encapsulation that produced $\mathbf{c}^*$; the attack would then be to query the decapsulation oracle on $\mathbf{c}$ $(\neq \mathbf{c}^*)$ to recover the encapsulated key.

At the same time, we stress that the above described attack is just a possibility and is not a *concrete* attack. Because it is quite possible that, by analyzing the decoding algorithm used in NTS-KEM decapsulation, one might show such invalid ciphertexts are computationally hard to generate adversarially.

A re-encoding step during NTS-KEM decapsulation, which is in line with the FO transformations, would entirely resolve this issue by correctly rejecting such invalid ciphertexts. Our proposed changes also perform the hash check $H_\ell(\mathbf{e}) = \mathbf{k}_e$ implicitly because of the following proposition shown in [ACP+19a].

Proposition 1 ([ACP+19a]). Let $\mathbf{c}^* = (\mathbf{c}_b \mid \mathbf{c}_c)$ be a correctly formed ciphertext for NTS-KEM with public key $\mathsf{pk} = (\mathbf{Q}, \tau, \ell)$. Then there exists a unique pair of vectors $((\mathbf{e}_a \mid \mathbf{r}_b), \mathbf{e})$ such that $\mathrm{hw}(\mathbf{e}) = \tau$ and $\mathbf{c}^* = (\mathbf{e}_a \mid \mathbf{r}_b) \cdot [\mathbf{I}_k \mid \mathbf{Q}] + \mathbf{e}$.

If a ciphertext $\mathbf{c}$ is not rejected by the new decapsulation oracle, it means that there is an error vector $\mathbf{e}$ with $\mathrm{hw}(\mathbf{e}) = \tau$ such that $\mathrm{Encode}(\mathsf{pk}, \mathbf{e}) = \mathbf{c}$. From Proposition 1, there then exists a unique pair of vectors $((\mathbf{e}_a \mid \mathbf{r}_b), \mathbf{e})$ w.r.t. $\mathbf{c}$, with $\mathrm{hw}(\mathbf{e}) = \tau$, such that $\mathbf{c} = (\mathbf{e}_a \mid \mathbf{r}_b) \cdot [\mathbf{I}_k \mid \mathbf{Q}] + \mathbf{e}$. It is clear that $\mathbf{r}_b = \mathbf{c}_b - \mathbf{e}_b$, and because of the uniqueness of $\mathbf{r}_b$, we must have $H_\ell(\mathbf{e}) = \mathbf{c}_b - \mathbf{e}_b$ in the evaluation of $\mathrm{Encode}(\mathsf{pk}, \mathbf{e})$. Because of this observation, our changes to NTS-KEM decapsulation also preserve the tightness of the initial IND-CCA security proof for NTS-KEM in the ROM, while fixing the flaw discussed above (refer to [ACP+19b] for more details on the updated ROM proof for NTS-KEM).

4 IND-CCA Security of NTS-KEM in the QROM

In this section, we will be providing a (game-hopping) security proof for NTS-KEM in the QROM, relying on techniques used in [JZC+18,SXY18]. Specifically, we show that NTS-KEM is IND-CCA secure in the QROM, if McEliece is OW secure as a PKE scheme.

Theorem 1. *In the quantum random oracle model, if there exists an adversary $\mathcal{A}$ winning the* IND-CCA *game for NTS-KEM with advantage ε, issuing at most q_D queries to the decapsulation oracle and at most q_H quantum queries to the random oracle $H_\ell(.)$, then there exists an adversary $\hat{\mathcal{B}}$ against the* OW *security of the McEliece PKE scheme with advantage at least $\frac{1}{4}\left(\frac{\varepsilon}{q_H} - \frac{1}{\sqrt{2^{\ell-1}}}\right)^2$, and the running time of $\hat{\mathcal{B}}$ is about that of $\mathcal{A}$.*

Similar to the IND-CCA security proof for NTS-KEM in the classical ROM given in [ACP+19a], we define NTS$^-$, a variant of NTS-KEM, which creates key encapsulations that are McEliece-type encryptions of message vectors of the form $\mathbf{m} = (\mathbf{e_a} \mid \mathbf{r_b})$, where $\mathbf{r_b} \leftarrow_\$ \mathbb{F}_2^\ell$, and $\mathbf{r_b}$ is considered to be the *encapsulated key* for NTS^-. This is in contrast to the original NTS-KEM scheme in which the encapsulations are encryptions of messages of the form $\mathbf{m} = (\mathbf{e_a} \mid \mathbf{k_e})$, where $\mathbf{k_e} = H_\ell(\mathbf{e}) \in \mathbb{F}_2^\ell$, and $\mathbf{K} = H_\ell(\mathbf{k_e} \mid \mathbf{e})$ is the encapsulated key. As will be seen later on, this step is convenient for our proof because we essentially *decouple* the need for random oracles from the NTS^- scheme.

Towards a reduction in the QROM from the IND-CCA security of NTS-KEM to the OW security of the McEliece PKE scheme, as an intermediate step we first note that NTS^- satisfies a non-standard security notion, denoted as *error one-wayness* or EOW security (as introduced in [ACP+19a]). This notion is specific to McEliece-type KEM schemes, e.g., NTS-KEM, which encrypt messages of the form $\mathbf{m} = (\mathbf{e_a} \mid \mathbf{r_b})$ with error vector $\mathbf{e} = (\mathbf{e_a} \mid \mathbf{e_b} \mid \mathbf{e_c})$ during key encapsulation. Roughly speaking, this notion states that it is hard to recover the error vector $\mathbf{e}$ used to generate a given challenge ciphertext $\mathbf{c}$. EOW security for NTS-KEM-like KEMs is defined formally in Fig. 4 where the adversary gets the encapsulation of a random key and is required to produce the error vector which led to that particular encapsulation.

$\text{EOW}_{\mathsf{KEM}}^{\mathcal{A}}$

$(\mathsf{pk}, \mathsf{sk}) \leftarrow \mathsf{KGen}(1^\lambda)$

$(\mathbf{K}, \mathbf{c}^*) \leftarrow \mathsf{Encap}(\mathsf{pk})$

$\mathbf{e}$: error vector used to produce $\mathbf{c}^*$

$\mathbf{e}' \leftarrow \mathcal{A}(1^\lambda, \mathsf{pk}, \mathbf{c}^*)$

return $(\mathbf{e}' = \mathbf{e})$

Fig. 4. EOW-security game for KEM.

Now NTS^- is EOW-secure because of the following security reduction shown in [ACP+19a], which does not rely on any random oracles.

Theorem 2 ([ACP+19a]). *If there is a (t, ε)-adversary $\mathcal{B}$ against the* EOW *security of NTS^-, then there is a (t, ε)-adversary $\hat{\mathcal{B}}$ against the* OW *security of the McEliece PKE scheme.*

So we can focus on reducing the IND-CCA security of NTS-KEM to the EOW security of NTS^- in the QROM.

Proof (of Theorem1). Let $\mathcal{A}$ be an adversary against the IND-CCA game for NTS-KEM with advantage ε, issuing at most q_D queries to the decapsulation oracle and at most q_H queries to the quantum random oracle $H_\ell(.)$.

Consider the games $G_0 - G_5$ described in Fig. 5. Here $\mathsf{pk} = (\mathbf{Q}, \tau, \ell)$ and $\mathsf{sk} = (\mathbf{a}^*, \mathbf{h}^*, \mathbf{p}, \mathbf{z}, \mathsf{pk})$ as described in Sect. 3 on NTS-KEM key generation. Also $H_0^+ : \{0,1\}^* \to \{0,1\}^\ell, H_1^n : \{0,1\}^n \to \{0,1\}^\ell, H_2^{\ell+n} : \{0,1\}^{\ell+n} \to \{0,1\}^\ell,$

Games $G_0 - G_5$

$b \leftarrow_{\$} \{0,1\}$
$(\mathsf{pk}, \mathsf{sk}) \leftarrow \mathsf{KGen}_{\text{NTS-KEM}}(1^\lambda)$
$\mathbf{e}^* \leftarrow_{\$} \{\mathbf{e} \in \mathbb{F}_2^n \mid \text{hw}(\mathbf{e}) = \tau\}$
$\mathbf{k}_\mathbf{e}^* = H_\ell(\mathbf{e}^*) /\!/ \ G_0\text{–}G_4; \ \mathbf{k}_\mathbf{e}^* \leftarrow_{\$} \mathbb{F}_2^\ell /\!/ \ G_5$
$(\mathbf{0_a} \mid \mathbf{c}^*) = (\mathbf{e}_\mathbf{a}^* \mid \mathbf{k}_\mathbf{e}^*) \cdot [\mathbf{I_k} \mid \mathbf{Q}] + \mathbf{e}^*$
$\mathbf{K}_0^* = H_\ell(\mathbf{k}_\mathbf{e}^* \mid \mathbf{e}^*) /\!/ \ G_0\text{–}G_4; \ \mathbf{K}_0^* \leftarrow_{\$} \mathbb{F}_2^\ell /\!/ \ G_5$
$\mathbf{K}_1^* \leftarrow_{\$} \mathbb{F}_2^\ell$
$b' \leftarrow \mathcal{A}^{H_\ell(.), \mathsf{Decap}(., sk)}(1^\lambda, \mathsf{pk}, \mathbf{K}_b^*, \mathbf{c}^*) /\!/ \ G_0\text{–}G_3$
$\ddot{H}_\ell(.) = H_\ell(.);$
$\quad \ddot{H}_\ell(\mathbf{e}^*) \leftarrow_{\$} \mathbb{F}_2^\ell; \ \ddot{H}_\ell(\mathbf{k}_\mathbf{e}^* \mid \mathbf{e}^*) \leftarrow_{\$} \mathbb{F}_2^\ell /\!/ \ G_4$
$b' \leftarrow \mathcal{A}^{\ddot{H}_\ell(.), \mathsf{Decap}(., sk)}(1^\lambda, \mathsf{pk}, \mathbf{K}_b^*, \mathbf{c}^*) /\!/ \ G_4$
return $(b' = b) /\!/ \ G_0\text{–}G_4$
$i \leftarrow_{\$} \{1, \ldots, q_H\} /\!/ \ G_5$
run $\mathcal{A}^{H_\ell(.), \mathsf{Decap}(., sk)}(1^\lambda, \mathsf{pk}, \mathbf{K}_b^*, \mathbf{c}^*)$ until
$\quad i$-th query to $(H_1^n \times [H_4^n \circ g])(.) /\!/ \ G_5$
measure the i-th query to be $\hat{\mathbf{e}} /\!/ \ G_5$
return $(\hat{\mathbf{e}} = \mathbf{e}^*) /\!/ \ G_5$

$H_\ell(\mathbf{x})$ // $|\mathbf{x}| \neq n, (\ell + n)$

return $H_0^+(\mathbf{x})$

$H_\ell(\mathbf{e})$ // $|\mathbf{e}| = n$

return $H_1^n(\mathbf{e})$

$H_\ell(\mathbf{k_e} \mid \mathbf{e})$ // $|(\mathbf{k_e} \mid \mathbf{e})| = (\ell + n)$

if $\text{hw}(\mathbf{e}) = \tau$ and
$\quad \mathbf{k_e} = H_\ell(\mathbf{e})$ **then** $/\!/ \ G_2\text{–}G_5$
$\quad \mathbf{c} = \text{Encode}(\mathsf{pk}, \mathbf{e}) /\!/ \ G_2\text{–}G_5$
$\quad$ **return** $H_4^n(\mathbf{1_a} \mid \mathbf{c}) /\!/ \ G_2\text{–}G_5$
return $H_2^{\ell+n}(\mathbf{k_e} \mid \mathbf{e})$

$\mathsf{Decap}(\mathbf{c} \neq \mathbf{c}^*, \mathsf{sk})$ // $G_0\text{–}G_2$

Parse $\mathsf{sk} = (\mathsf{sk}', \mathbf{z})$
$\mathbf{e} = \text{Decode}(\mathsf{sk}', \mathbf{c})$
$\mathbf{k_e} = \mathbf{c_b} - \mathbf{e_b}$
if $\text{hw}(\mathbf{e}) = \tau$ and $\text{Encode}(\mathsf{pk}, \mathbf{e}) = \mathbf{c}$ **then**
$\quad$ **return** $\mathbf{K} = H_\ell(\mathbf{k_e} \mid \mathbf{e})$
else return
$\quad \mathbf{K} = H_\ell(\mathbf{z} \mid \mathbf{1_a} \mid \mathbf{c}) /\!/ \ G_0$
$\quad \mathbf{K} = H_3^n(\mathbf{1_a} \mid \mathbf{c}) /\!/ \ G_1\text{–}G_2$

$\mathsf{Decap}(\mathbf{c} \neq \mathbf{c}^*, \mathsf{sk})$ // $G_3\text{–}G_5$

return $\mathbf{K} = H_4^n(\mathbf{1_a} \mid \mathbf{c})$

Fig. 5. Games G_0–G_5 for the proof of Theorem 1.

$H_3^n : \{0,1\}^n \rightarrow \{0,1\}^\ell$ and $H_4^n : \{0,1\}^n \rightarrow \{0,1\}^\ell$ are independent random functions that are used in the evaluation of queries (of varying lengths) w.r.t. the oracles $H_\ell(.)$ and $\mathsf{Decap}(., \mathsf{sk})$ in the games. $\text{Decode}(\mathsf{sk}', \mathbf{c})$ defined over ciphertexts $\mathbf{c} \in \mathbb{F}_2^{n-k+\ell}$ recovers an error vector after applying the decoding algorithm used in NTS-KEM decapsulation [*Decapsulation*, Sect. 3] and the permutation $\mathbf{p}$ of the secret key.

Game G_0. The game G_0 is exactly the IND-CCA game for NTS-KEM. So,

$$|\Pr[G_0^\mathcal{A} \implies 1] - \frac{1}{2}| = \varepsilon$$

where "$\mathrm{G}_i^{\mathcal{A}} \implies 1$" denotes the event that the game G_i returns 1 w.r.t. the adversary $\mathcal{A}$.

Game G_1. In game G_1, we modify the decapsulation oracle such that $H_3^n(\mathbf{1}_a \mid \mathbf{c})$ is returned instead of $H_\ell(\mathbf{z} \mid \mathbf{1}_a \mid \mathbf{c})$ for an invalid ciphertext $\mathbf{c}$, i.e., pseudo-random decapsulations of invalid ciphertexts are replaced by truly random outputs. We use Lemma 2 to claim that there is a negligible difference between $\mathcal{A}$'s winning probabilities in games G_0 and G_1, i.e.,

$$|\Pr[\mathrm{G}_1^{\mathcal{A}} \implies 1] - \Pr[\mathrm{G}_0^{\mathcal{A}} \implies 1]| \leq q_H \cdot 2^{-\frac{\ell+1}{2}}$$

The proof for this claim follows along similar lines to that of [SXY18, Lemma 2.2], but with modifications to account for the specific way NTS-KEM rejects invalid ciphertexts during decapsulation. To prove our claim, we again consider the following sequence of games for $\mathcal{A}$ based on the random oracles it has access to, that are relevant during the transition from G_0 to G_1.

G_0: The game returns accordingly as $\mathcal{A}^{H_2^{\ell+n}(.),H_2^{\ell+n}(\mathbf{z}|\mathbf{1_a}|.)}(.)$ outputs, where $\mathbf{z} \leftarrow_\$ \{0,1\}^\ell$ is a part of the secret key sk.

$\mathrm{G}_{0.5}$: The game returns accordingly as $\mathcal{A}^{O^{\ell+n}[\mathbf{z},\mathbf{a},H_2^{\ell+n},H_3^n](.),H_3^n(\mathbf{1_a}|.)}(.)$ outputs, where $O^{\ell+n}[\mathbf{z},\mathbf{a},H_2^{\ell+n},H_3^n](.)$ is a function defined as

$$O^{\ell+n}[\mathbf{z},\mathbf{a},H_2^{\ell+n},H_3^n](\mathbf{z}' \mid \mathbf{c}) = \begin{cases} H_2^{\ell+n}(\mathbf{z}' \mid \mathbf{c}) & \text{if } \mathbf{z}' \neq \mathbf{z} \text{ or } [\mathbf{c}]_a \neq \mathbf{1_a} \\ H_3^n(\mathbf{c}) & \text{otherwise} \end{cases}$$

Here, $[x]_b$ denotes the first b-bits of input x.

G_1: The game returns accordingly as $\mathcal{A}^{H_2^{\ell+n}(.),H_3^n(\mathbf{1_a}|.)}(.)$ outputs.

Note that $\Pr[\mathrm{G}_{0.5}^{\mathcal{A}} \implies 1] = \Pr[\mathrm{G}_0^{\mathcal{A}} \implies 1]$: for $(\ell+n)$-bit queries of the form $(\mathbf{z} \mid \mathbf{1_a} \mid .)$, the function $O^{\ell+n}[\mathbf{z},\mathbf{a},H_2^{\ell+n},H_3^n](.)$ makes sure that we maintain consistency of the oracle evaluations when replacing $H_2^{\ell+n}(\mathbf{z} \mid \mathbf{1_a} \mid .)$ with $H_3^n(\mathbf{1_a} \mid .)$.

We show that $|\Pr[\mathrm{G}_1^{\mathcal{A}} \implies 1] - \Pr[\mathrm{G}_{0.5}^{\mathcal{A}} \implies 1]| \leq q_H \cdot 2^{-\frac{\ell+1}{2}}$ via a reduction to Lemma 2. Consider the algorithm $\mathcal{C}$ that has oracle access to the function $g(.)$ which is either $g_{\mathbf{z}}(.)$ for uniformly random $\mathbf{z} \leftarrow_\$ \{0,1\}^\ell$ or $g_\perp(.)$, where the functions $g_{\mathbf{z}}(.)$ and $g_\perp(.)$ are as defined in Lemma 2. $\mathcal{C}^{g(.)}$ runs $\mathcal{A}^{\hat{O}^{\ell+n}(.),H_3^n(\mathbf{1_a}|.)}(.)$ where $\mathcal{C}$ simulates the oracles $H_2^{\ell+n}(.)$ and $H_3^n(.)$ using two different $2q_H$-wise independent functions respectively (see Lemma 1), and simulates $\hat{O}^{\ell+n}(.)$ as follows: When $\mathcal{A}$ queries $(\mathbf{z}' \mid \mathbf{c})$ to $\hat{O}^{\ell+n}(.)$, $\mathcal{B}$ queries $\mathbf{z}'$ to $g(.)$ and gets a bit b. If $b = 1$ and $[\mathbf{c}]_a = \mathbf{1_a}$, then $\mathcal{C}$ returns $H_3^n(\mathbf{c})$. Otherwise, $\mathcal{C}$ returns $H_2^{\ell+n}(\mathbf{z}' \mid \mathbf{c})$.

It is clear that if $g(.) = g_{\mathbf{z}}(.)$ for uniformly random $\mathbf{z} \leftarrow_\$ \{0,1\}^\ell$, $\mathcal{C}$ perfectly simulates $\mathrm{G}_{0.5}$ in $\mathcal{A}$'s view, and similarly if $g(.) = g_\perp(.)$, $\mathcal{C}$ simulates G_1. Thus, we get

$$|\Pr[\mathrm{G}_1^{\mathcal{A}} \implies 1] - \Pr[\mathrm{G}_{0.5}^{\mathcal{A}} \implies 1]| = |\Pr[\mathcal{C}^{g_\perp(.)}(.) \to 1] - \Pr[\mathcal{C}^{g_{\mathbf{z}}(.)}(.) \to 1 \mid \mathbf{z} \leftarrow_\$ \{0,1\}^\ell]|$$

Since the number of $\mathcal{C}$'s oracle queries to $g(.)$ is the same as the number of $\mathcal{A}$'s queries to $\hat{O}^{\ell+n}(.)$, we can use Lemma 2 to further obtain

$$|\Pr[\mathcal{C}^{g_\perp(.)}(.) \to 1] - \Pr[\mathcal{C}^{g_{\mathbf{z}}(.)}(.) \to 1 \mid \mathbf{z} \leftarrow_\$ \{0,1\}^\ell]| \leq q_H \cdot 2^{-\frac{\ell+1}{2}}$$

which proves our claim regarding the adversary $\mathcal{A}$'s winning probabilities in games G_0 and G_1.

Game G_2. In game G_2, the encapsulated keys are derived in a different way: if the $(\ell+n)$-bit input $(\mathbf{k_e} \mid \mathbf{e})$ to $H_\ell(.)$ is of the *correct* form, i.e., $\text{hw}(\mathbf{e}) = \tau$ and $\mathbf{k_e} = H_\ell(\mathbf{e})$, then the output is replaced by $H_4^n(\mathbf{1_a} \mid \mathbf{c})$, where $\mathbf{c} = \text{Encode}(\mathsf{pk}, \mathbf{e})$.

Because NTS-KEM is a *perfectly correct* scheme as shown in [ACP+19a] (see Subsect. 2.3 for correctness definition), we note that $\text{Encode}(\mathsf{pk}, .)$ is injective, and thus, $H_4^n(\mathbf{1_a} \mid \text{Encode}(\mathsf{pk}, .))$ returns perfectly random values for distinct inputs of the type $(H_\ell(\mathbf{e}) \mid \mathbf{e})$ with $\text{hw}(\mathbf{e}) = \tau$. As the oracle distributions of $H_\ell(.)$ are equivalent in games G_1 and G_2, we have

$$\Pr[G_2^{\mathcal{A}} \implies 1] = \Pr[G_1^{\mathcal{A}} \implies 1]$$

Game G_3. In game G_3, we change the decapsulation oracle such that there is no need for the secret key sk. Specifically, when the adversary $\mathcal{A}$ asks for the decapsulation of a ciphertext $\mathbf{c}$ ($\neq \mathbf{c}^*$, the challenge ciphertext), $H_4^n(\mathbf{1_a} \mid \mathbf{c})$ is returned. Let $\mathbf{e} = \text{Decode}(\mathsf{sk}', \mathbf{c})$ and $\mathbf{k_e} = \mathbf{c_b} - \mathbf{e_b}$. Consider the following two cases:

Case 1: If the checks in NTS-KEM decapsulation – i.e., $\text{hw}(\mathbf{e}) = \tau$ and $\text{Encode}(\mathsf{pk}, \mathbf{e}) = \mathbf{c}$ – are satisfied, then the decapsulation oracles in games G_2 and G_3 return $H_\ell(\mathbf{k}_e \mid \mathbf{e})$ and $H_4^n(\mathbf{1_a} \mid \mathbf{c})$ respectively. In G_2, as discussed in Subsect. 3.1, the re-encoding step does an implicit hash check, and hence, we also have $H_\ell(\mathbf{e}) = \mathbf{c}_b - \mathbf{e}_b = \mathbf{k}_e$. Therefore, $H_\ell(\mathbf{k}_e \mid \mathbf{e})$ evaluates to $H_4^n(\mathbf{1_a} \mid \text{Encode}(\mathsf{pk}, \mathbf{e})) = H_4^n(\mathbf{1_a} \mid \mathbf{c})$ in G_2, which is the value returned in G_3 as well.

Case 2: If one of the checks is not satisfied, then the values $H_3^n(\mathbf{1_a} \mid \mathbf{c})$ and $H_4^n(\mathbf{1_a} \mid \mathbf{c})$ are returned in games G_2 and G_3 respectively. In G_2, the function $H_3^n(.)$ is independent of all other random oracles, and thus, the output $H_3^n(\mathbf{1_a} \mid \mathbf{c})$ is uniformly random in $\mathcal{A}$'s view. In G_3, the only way $\mathcal{A}$ gets prior access to the oracle $H_4^n(.)$ is if it already queried $H_\ell(.)$ with an input of the type $(\mathbf{k}'_\mathbf{e} \mid \mathbf{e}')$ such that $\text{hw}(\mathbf{e}') = \tau$ and $\mathbf{k}'_\mathbf{e} = H_\ell(\mathbf{e}')$, and got back $H_4^n(\mathbf{1_a} \mid \text{Encode}(\mathsf{pk}, \mathbf{e}'))$. Now it's not hard to see that $\text{Encode}(\mathsf{pk}, \mathbf{e}')$ cannot be equal to $\mathbf{c}$,[7] which implies that the output of the modified decapsulation oracle $H_4^n(\mathbf{1_a} \mid \mathbf{c})$ is a *fresh* random value like $H_3^n(\mathbf{1_a} \mid \mathbf{c})$.

Because the output distributions of the decapsulation oracles in games G_2 and G_3 are the same in both cases, we have

$$\Pr[G_3^{\mathcal{A}} \implies 1] = \Pr[G_2^{\mathcal{A}} \implies 1]$$

Game G_4. In game G_4, we *reprogram* the random oracle $H_\ell(.)$ on inputs $\mathbf{e}^*$ and $(\mathbf{k}_\mathbf{e}^* \mid \mathbf{e}^*)$ such that they result in fresh uniformly random outputs. To be specific, we replace $H_\ell(.)$ with the function $\ddot{H}_\ell(.)$ where $\ddot{H}_\ell(\mathbf{e}^*) = \mathbf{r}_\mathbf{b}^* \leftarrow_\$ \mathbb{F}_2^\ell$ and $\ddot{H}_\ell(\mathbf{k}_\mathbf{e}^* \mid \mathbf{e}^*) = \dot{\mathbf{K}}_\mathbf{0}^* \leftarrow_\$ \mathbb{F}_2^\ell$, and $\ddot{H}_\ell(.) = H_\ell(.)$ everywhere else. It is clear that in

[7] On the contrary, if $\text{Encode}(\mathsf{pk}, \mathbf{e}') = \mathbf{c}$, then because of NTS-KEM correctness, we have $\text{Decode}(\mathsf{sk}', \mathbf{c}) = \mathbf{e}' = \mathbf{e}$. This means that the checks $\text{hw}(\mathbf{e}) = \tau$ and $\text{Encode}(\mathsf{pk}, \mathbf{e}) = \mathbf{c}$ are satisfied, a contradiction.

this game, as we are *masking* the information used to derive the challenge pair $(\mathbf{K}_b^*, \mathbf{c}^*)$ from $\mathcal{A}$'s view, its output is independent of bit b. Therefore,

$$\Pr[\mathrm{G}_4^{\mathcal{A}} \implies 1] = \frac{1}{2}$$

In order to bound the difference in $\mathcal{A}$'s winning probabilities in games G_3 and G_4, we use Lemma 3. Let the function $g(.)$ defined over error vectors $\mathbf{e} \in \mathbb{F}_2^n$ be as follows,

$$g(\mathbf{e}) = \begin{cases} \mathbf{1_a} \mid \mathrm{Encode}(\mathsf{pk}, \mathbf{e}) & \text{if } \mathrm{hw}(\mathbf{e}) = \tau \\ \mathbf{0_a} \mid \mathbf{e_b} \mid \mathbf{e_c} & \text{if } \mathrm{hw}(\mathbf{e}) \neq \tau \end{cases}$$

Looking at game G_3, the oracle query $H_\ell(\mathbf{k_e^*} \mid \mathbf{e}^*)$ evaluates to $H_4^n(g(\mathbf{e}^*)) = H_4^n \circ g\,(\mathbf{e}^*)$. So we are actually reprogramming the oracles $H_1^n(.)$ and $H_4^n \circ g\,(.)$ at the input $\mathbf{e}^*$.

Define the function $\ddot{H}_4^n \circ g\,(.)$ such that $\ddot{H}_4^n \circ g\,(\mathbf{e}^*) \leftarrow_\$ \mathbb{F}_2^\ell$ and $\ddot{H}_4^n \circ g\,(.) = H_4^n \circ g\,(.)$ everywhere else. Similarly, let $\ddot{H}_1^n(\mathbf{e}^*) \leftarrow_\$ \mathbb{F}_2^\ell$ and $\ddot{H}_1^n(.) = H_1^n(.)$ everywhere else. Now let the oracles $(H_1^n \times [H_4^n \circ g])(.) = (H_1^n(.), H_4^n \circ g\,(.))$[8] and $(\ddot{H}_1^n \times [\ddot{H}_4^n \circ g])(.) = (\ddot{H}_1^n(.), \ddot{H}_4^n \circ g\,(.))$. If we also have a function $\hat{H}_4^n(.)$ such that $\hat{H}_4^n(g(\mathbf{e}^*)) = \perp$ and $\hat{H}_4^n(.) = H_4^n(.)$ everywhere else, then $\hat{H}_4^n(\mathbf{1_a} \mid .)$ is precisely the (unchanged) decapsulation oracle in games G_3 and G_4.

Let $\mathcal{U}^{(H_1^n \times [H_4^n \circ g]), \hat{H}_4^n}$ be an algorithm described in Fig. 6 that has quantum access to the oracles $(H_1^n \times [H_4^n \circ g])(.)$ and $\hat{H}_4^n(.)$, and takes an input $(\mathsf{pk}, \mathbf{e}^*, (\mathbf{k_e^*}, \mathbf{K_0^*}))$ which is derived in the same way as in games G_3 and G_4; i.e., $(\mathsf{pk}, \mathsf{sk}) \leftarrow \mathsf{KGen}_{\text{NTS-KEM}}(1^\lambda)$, $\mathbf{e}^* \leftarrow_\$ \{\mathbf{e} \in \mathbb{F}_2^n \mid \mathrm{hw}(\mathbf{e}) = \tau\}, \mathbf{k_e^*} = H_1^n(\mathbf{e}^*)$ and $\mathbf{K_0^*} = H_4^n \circ g(\mathbf{e}^*)$, with functions $H_1^n(.), H_4^n(.)$ and $g(.)$ as previously described.

Here the random functions $H_0^+ : \{0,1\}^* \to \{0,1\}^\ell$ and $H_2^{\ell+n} : \{0,1\}^{\ell+n} \to \{0,1\}^\ell$ are independently sampled by the algorithm. Note that, $\mathcal{U}^{(H_1^n \times [H_4^n \circ g]), \hat{H}_4^n}$ on input $(\mathsf{pk}, \mathbf{e}^*, (\mathbf{k_e^*}, \mathbf{K_0^*}))$ simulates G_3 in the adversary $\mathcal{A}$'s view, whereas the algorithm $\mathcal{U}^{(\ddot{H}_1^n \times [\ddot{H}_4^n \circ g]), \hat{H}_4^n}$ on the same input $(\mathsf{pk}, \mathbf{e}^*, (\mathbf{k_e^*}, \mathbf{K_0^*}))$ simulates G_4. Also $\mathcal{A}$ can have (separate) access to the *internal* oracles $H_1^n(.)$ and $[H_4^n \circ g](.)$ by querying $H_\ell(.)$, which could be simulated by $\mathcal{U}$ by accessing $(H_1^n \times [H_4^n \circ g])(.)$ and ignoring part of the output of the oracle using a trick[9] described in [BZ13, TU16]. Therefore, the number of oracle queries to $(H_1^n \times [H_4^n \circ g])(.)$ is at most q_H.

Let $\mathcal{V}^{(\ddot{H}_1^n \times [\ddot{H}_4^n \circ g]), \hat{H}_4^n}$ be an algorithm that on input $(\mathsf{pk}, \mathbf{e}^*, (\mathbf{k_e^*}, \mathbf{K_0^*}))$ does the following: samples $i \leftarrow_\$ \{1, \ldots, q_H\}$, runs $\mathcal{U}^{(\ddot{H}_1^n \times [\ddot{H}_4^n \circ g]), \hat{H}_4^n}$ until the i-th query to $(\ddot{H}_1^n \times [\ddot{H}_4^n \circ g])(.)$ and returns the measurement outcome of the query in the computational basis (if $\mathcal{U}$ makes less than i queries, the algorithm outputs $\perp$).

[8] For error vectors $\mathbf{e} \in \mathbb{F}_2^n$ with $\mathrm{hw}(\mathbf{e}) \neq \tau$, the reason we defined $g(\mathbf{e})$ – even though $\mathcal{A}$ only has access to $H_4^n(\mathbf{1_a} \mid .)$ in games G_3 and G_4 – is to have a consistent domain $(\mathbb{F}_2^n)$ and co-domain $(\mathbb{F}_2^\ell)$ between the oracles $H_1^n(.)$ and $H_4^n \circ g\,(.)$. This would be helpful, for example, when applying Lemma 3 in our setting.

[9] For example, if we want to access $H_1^n(.)$ by making queries to $(H_1^n \times [H_4^n \circ g])(.)$, then we just have to prepare a uniform superposition of all states in the output register corresponding to $H_4^n \circ g(.)$.

$\mathcal{U}^{(H_1^n \times [H_4^n \circ g]), \hat{H}_4^n}(\mathsf{pk}, \mathbf{e}^*, (\mathbf{k}_\mathbf{e}^*, \mathbf{K}_\mathbf{0}^*))$

$\mathbf{c}^* = \text{Encode}(\mathsf{pk}, \mathbf{e}^*)$

$\mathbf{K}_1^* \leftarrow_\$ \mathbb{F}_2^\ell$

$b \leftarrow_\$ \{0, 1\}$

$b' \leftarrow \mathcal{A}^{H_\ell(.), \mathsf{Decap}(., sk)}(1^\lambda, \mathsf{pk}, \mathbf{K}_b^*, \mathbf{c}^*)$

return $(b' = b)$

$\mathsf{Decap}(\mathbf{c} \neq \mathbf{c}^*, \mathsf{sk})$

return $K = \hat{H}_4^n(\mathbf{1_a} \mid \mathbf{c})$

$H_\ell(\mathbf{x})$ // $|\mathbf{x}| \neq n, (\ell + n)$

return $H_0^+(\mathbf{x})$

$H_\ell(\mathbf{e})$ // $|\mathbf{e}| = n$

return $H_1^n(\mathbf{e})$

$H_\ell(\mathbf{k_e} \mid \mathbf{e})$// $|(\mathbf{k_e} \mid \mathbf{e})| = (\ell + n)$

if $\mathbf{e} = \mathbf{e}^*$ **then**

  if $\mathbf{k_e} = \mathbf{k_e^*}$ **then**

    return $H_4^n \circ g(\mathbf{e})$

  else return $H_2^{\ell+n}(\mathbf{k_e} \mid \mathbf{e})$

elseif $\mathrm{hw}(\mathbf{e}) = \tau$ and

  $\mathbf{k_e} = H_\ell(\mathbf{e})$ **then**

  return $H_4^n \circ g(\mathbf{e})$

else return $H_2^{\ell+n}(\mathbf{k_e} \mid \mathbf{e})$

Fig. 6. Algorithm $\mathcal{U}^{(H_1^n \times [H_4^n \circ g]), \hat{H}_4^n}$ for the proof of Theorem 1.

Game G$_5$. From the description of G$_5$, we see that $\Pr[\mathcal{V}^{(\ddot{H}_1^n \times [\ddot{H}_4^n \circ g]), \hat{H}_4^n} \Longrightarrow \mathbf{e}^*] = \Pr[G_5^\mathcal{A} \Longrightarrow 1]$ because, as previously discussed regarding the winning probability in G$_4$, the oracle $(\ddot{H}_1^n \times [\ddot{H}_4^n \circ g])(.)$ does not reveal any information about $H_\ell(\mathbf{e}^*)$ and $H_\ell(\mathbf{k_e^*} \mid \mathbf{e}^*)$. So applying Lemma 3 with the setting $\mathcal{O}_1 = (H_1^n \times [H_4^n \circ g])(.), \ddot{\mathcal{O}}_1 = (\ddot{H}_1^n \times [\ddot{H}_4^n \circ g])(.), \mathcal{O}_2 = \hat{H}_4^n(\mathbf{1_a} \mid .), inp = \mathsf{pk}, x = \mathbf{e}^*$ and $y = (\mathbf{k_e^*}, \mathbf{K_0^*})$, we obtain[10],

$$|\Pr[G_4^\mathcal{A} \Longrightarrow 1] - \Pr[G_3^\mathcal{A} \Longrightarrow 1]| \leq 2q_H \sqrt{\Pr[G_5^\mathcal{A} \Longrightarrow 1]}$$

Finally, we construct an adversary $\mathcal{B}$ against the EOW security of NTS$^-$ (as described in Fig. 4) such that its advantage is $\Pr[G_5^\mathcal{A} \Longrightarrow 1]$. Given an input $(1^\lambda, \mathsf{pk}, \mathbf{c}^*)$, $\mathcal{B}$ does the following:

- Runs $\mathcal{A}$ as a subroutine in game G$_5$.
- Uses four different $2q_H$-wise independent functions to *perfectly* simulate the random oracles $H_0^+(.), H_1^n(.), H_4^n(.)$ and $H_2^{l+n}(.)$ respectively in $\mathcal{A}$'s view, as described in Lemma 1. Also evaluates $H_1^n(.)$ and $[H_4^n \circ g](.)$ at $\mathcal{A}$'s queries using the oracle $(H_1^n \times [H_4^n \circ g])(.)$.
- Answers decapsulation queries using the function $H_4^n(\mathbf{1_a} \mid .)$.
- For $\mathcal{A}$'s challenge query, it samples $\mathbf{K}^* \leftarrow_\$ \mathbb{F}_2^\ell$ and responds with $(\mathbf{K}^*, \mathbf{c}^*)$.

[10] The original OW2H lemma of [Unr14] would have required $\mathbf{e}^*$ to be sampled uniformly in $\mathbb{F}_2^n$, the domain of $(H_1^n \times [H_4^n \circ g])(.)$. Therefore we use Lemma 3 which generalizes to arbitrary distributions of $\mathbf{e}^*$; in particular, $\mathbf{e}^* \leftarrow_\$ \{\mathbf{e} \in \mathbb{F}_2^n \mid \mathrm{hw}(\mathbf{e}) = \tau\}$.

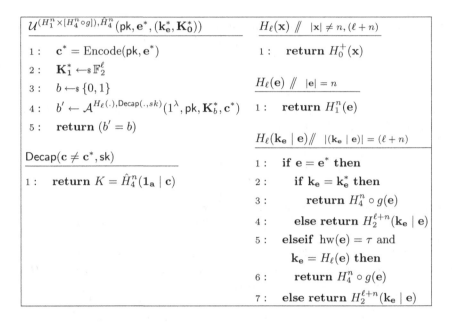

– Samples $i \leftarrow_{\$} \{1, \ldots, q_H\}$, measures the i-th query to the oracle $(H_1^n \times [H_4^n \circ g])(.)$ and returns the outcome $\hat{\mathbf{e}}$.

From the above description of $\mathcal{B}$, we note that its EOW advantage against NTS$^-$ is indeed $\Pr[\mathrm{G}_5^{\mathcal{A}} \implies 1]$. Coming to the running times of $\mathcal{A}$ and $\mathcal{B}$, say $t_{\mathcal{A}}$ and $t_{\mathcal{B}}$ respectively, if t_{Enc} denotes the time needed to perform a single *Encode*(pk, .) operation, we have $t_{\mathcal{B}} \approx t_{\mathcal{A}} + (q_H + q_D) \cdot O(q_H) + q_H \cdot t_{\mathrm{Enc}}$, i.e., the overhead is due to the simulation of $H_\ell(.)$ and $\mathsf{Decap}(., sk)$ oracles by $\mathcal{B}$.

By combining the bounds obtained w.r.t. the winning probabilities of $\mathcal{A}$ in each of the previous games and applying the security reduction of Theorem 2 to the EOW adversary $\mathcal{B}$, we obtain an adversary $\hat{\mathcal{B}}$ against the OW security of the McEliece PKE scheme with a running time ($= t_{\mathcal{B}}$) close to that of $\mathcal{A}$, and advantage $\hat{\varepsilon}$ where

$$\hat{\varepsilon} \geq \frac{1}{4}\Big(\frac{\varepsilon}{q_H} - \frac{1}{\sqrt{2^{\ell-1}}}\Big)^2 \tag{1}$$

□

5 Conclusion

In this paper, we analyzed the security of NTS-KEM – a second round PKE/KEM candidate in NIST's PQC standardization project – in the QROM. Specifically, we identified an issue in the IND-CCA security proof of NTS-KEM in the classical ROM and suggested some modifications to the scheme towards fixing it. We later showed that our changes not only preserve the tightness of the intended ROM proof for NTS-KEM but also lead to an IND-CCA security reduction in the QROM. The proposed changes were later adopted by the NTS-KEM team in an update to their second round submission [ACP+19b].

We also note that our QROM reduction can be made tighter by using newer OW2H lemmas of [AHU19] and [BHH+19]. For example, one could consider the improved security reduction of the $\mathrm{U}_m^{\not\perp}$ transform in [BHH+19] to get rid of the factor $1/q_H$ from the term "ε/q_H" in Eq. (1). However, the quadratic loss in degree of tightness incurred by our reduction might still be unavoidable in the QROM [JZM19].

Acknowledgments. It is my pleasure to thank Kenny Paterson, and the rest of the NTS-KEM team, for helpful discussions. I would also like to thank the anonymous reviewers of CBCrypto 2020 for their comments.

References

[ACP+19a] Albrecht, M., Cid, C., Paterson, K.G., Tjhai, C.J., Tomlinson, M.: NTS-KEM: NIST PQC Second Round Submission (2019). https://nts-kem.io/

[ACP+19b] Albrecht, M., Cid, C., Paterson, K.G., Tjhai, C.J., Tomlinson, M.: NTS-KEM: NIST PQC Updated Second Round Submission (2019). https://nts-kem.io/

[AHU19] Ambainis, A., Hamburg, M., Unruh, D.: Quantum security proofs using semi-classical oracles. In: Boldyreva, A., Micciancio, D. (eds.) CRYPTO 2019, Part II. LNCS, vol. 11693, pp. 269–295. Springer, Cham (2019). https://doi.org/10.1007/978-3-030-26951-7_10

[BCNP08] Boyd, C., Cliff, Y., Gonzalez Nieto, J., Paterson, K.G.: Efficient one-round key exchange in the standard model. In: Mu, Y., Susilo, W., Seberry, J. (eds.) ACISP 2008. LNCS, vol. 5107, pp. 69–83. Springer, Heidelberg (2008). https://doi.org/10.1007/978-3-540-70500-0_6

[BDF+11] Boneh, D., Dagdelen, Ö., Fischlin, M., Lehmann, A., Schaffner, C., Zhandry, M.: Random oracles in a quantum world. In: Lee, D.H., Wang, X. (eds.) ASIACRYPT 2011. LNCS, vol. 7073, pp. 41–69. Springer,pg Heidelberg (2011). https://doi.org/10.1007/978-3-642-25385-0_3

[BHH+19] Bindel, N., Hamburg, M., Hövelmanns, K., Hülsing, A., Persichetti, E.: Tighter proofs of CCA security in the quantum random oracle model. In: Hofheinz, D., Rosen, A. (eds.) TCC 2019, Part II. LNCS, vol. 11892, pp. 61–90. Springer, Cham (2019). https://doi.org/10.1007/978-3-030-36033-7_3

[BR93] Bellare, M., Rogaway, P.: Random oracles are practical: a paradigm for designing efficient protocols. In: Denning, D.E., Pyle, R., Ganesan, R., Sandhu, R.S., Ashby, V. (eds.) ACM CCS 93: 1st Conference on Computer and Communications Security, Fairfax, Virginia, pp. 62–73. ACM Press (1993)

[BZ13] Boneh, D., Zhandry, M.: Secure signatures and chosen ciphertext security in a quantum computing world. In: Canetti, R., Garay, J.A. (eds.) CRYPTO 2013, Part II. LNCS, vol. 8043, pp. 361–379. Springer, Heidelberg (2013). https://doi.org/10.1007/978-3-642-40084-1_21

[CS03] Cramer, R., Shoup, V.: Design and analysis of practical public-key encryption schemes secure against adaptive chosen ciphertext attack. SIAM J. Comput. **33**(1), 167–226 (2003)

[Den03] Dent, A.W.: A designer s guide to KEMs. In: Paterson, K.G. (ed.) Cryptography and Coding 2003. LNCS, vol. 2898, pp. 133–151. Springer, Heidelberg (2003). https://doi.org/10.1007/978-3-540-40974-8_12

[FO13] Fujisaki, E., Okamoto, T.: Secure integration of asymmetric and symmetric encryption schemes. J. Cryptol. **26**(1), 80–101 (2013). https://doi.org/10.1007/s00145-011-9114-1

[FOPS01] Fujisaki, E., Okamoto, T., Pointcheval, D., Stern, J.: RSA-OAEP is secure under the RSA assumption. In: Kilian, J. (ed.) CRYPTO 2001. LNCS, vol. 2139, pp. 260–274. Springer, Heidelberg (2001). https://doi.org/10.1007/3-540-44647-8_16

[HHK17] Hofheinz, D., Hövelmanns, K., Kiltz, E.: A modular analysis of the Fujisaki-Okamoto transformation. In: Kalai, Y., Reyzin, L. (eds.) TCC 2017, Part I. LNCS, vol. 10677, pp. 341–371. Springer, Cham (2017). https://doi.org/10.1007/978-3-319-70500-2_12

[JZC+18] Jiang, H., Zhang, Z., Chen, L., Wang, H., Ma, Z.: IND-CCA-Secure key encapsulation mechanism in the quantum random oracle model, revisited. In: Shacham, H., Boldyreva, A. (eds.) CRYPTO 2018, Part III. LNCS, vol. 10993, pp. 96–125. Springer, Cham (2018). https://doi.org/10.1007/978-3-319-96878-0_4

[JZM19] Jiang, H., Zhang, Z., Ma, Z.: On the non-tightness of measurement-based reductions for key encapsulation mechanism in the quantum random oracle model. IACR Cryptology ePrint Archive 2019/494 (2019). https://eprint.iacr.org/2019/494

[McE78] McEliece, R.J.: A public-key cryptosystem based on algebraic coding theory. Deep Space Netw. Prog. Rep. **44**, 114–116 (1978)

[Nie86] Niederreiter, H.: Knapsack-type cryptosystems and algebraic coding theory. Probl. Control Inf. Theory **15**, 159–166 (1986)

[NIS15] NIST. FIPS PUB 202 Federal Information Processing Standards Publication: SHA-3 Standard: Permutation-Based Hash and Extendable-Output Functions, August 2015

[NIS19] NIST. Post-Quantum Cryptography Standardization: Second Round Submissions, January 2019. https://csrc.nist.gov/projects/post-quantum-cryptography/round-2-submissions

[SXY18] Saito, T., Xagawa, K., Yamakawa, T.: Tightly-secure key-encapsulation mechanism in the quantum random oracle model. In: Nielsen, J.B., Rijmen, V. (eds.) EUROCRYPT 2018, Part III. LNCS, vol. 10822, pp. 520–551. Springer, Cham (2018). https://doi.org/10.1007/978-3-319-78372-7_17

[SY17] Song, F., Yun, A.: Quantum security of NMAC and related constructions. In: Katz, J., Shacham, H. (eds.) CRYPTO 2017, Part II. LNCS, vol. 10402, pp. 283–309. Springer, Cham (2017). https://doi.org/10.1007/978-3-319-63715-0_10

[TU16] Targhi, E.E., Unruh, D.: Post-quantum security of the Fujisaki-Okamoto and OAEP transforms. In: Hirt, M., Smith, A. (eds.) TCC 2016, Part II. LNCS, vol. 9986, pp. 192–216. Springer, Heidelberg (2016). https://doi.org/10.1007/978-3-662-53644-5_8

[Unr14] Unruh, D.: Revocable quantum timed-release encryption. In: Nguyen, P.Q., Oswald, E. (eds.) EUROCRYPT 2014. LNCS, vol. 8441, pp. 129–146. Springer, Heidelberg (2014). https://doi.org/10.1007/978-3-642-55220-5_8

[Zha12] Zhandry, M.: Secure identity-based encryption in the quantum random oracle model. In: Safavi-Naini, R., Canetti, R. (eds.) CRYPTO 2012. LNCS, vol. 7417, pp. 758–775. Springer, Heidelberg (2012). https://doi.org/10.1007/978-3-642-32009-5_44

On the Decipherment of Sidel'nikov-Type Cryptosystems

Vladimir M. Deundyak[1,2], Yury V. Kosolapov[1(✉)], and Igor A. Maystrenko[1]

[1] Southern Federal University, B. Sadovaya, 105/42, 344006 Rostov-on-Don, Russia
{vmdeundyak,yvkosolapov}@sfedu.ru, mr.igmai@mail.ru
[2] FSASI RI Specvuzavtomatika, Gazetny, 51, 344002 Rostov-on-Don, Russia

Abstract. Asymmetric Sidel'nikov-type code-based cryptosystem is one of the generalizations of McEliece-type cryptosystems. The secret key in this generalization is the set of u randomly selected generator matrices G_{C_1},...,G_{C_u} of different $[n_i, k]$-codes C_i, $i \in \{1, ..., u\}$. In addition, a part of the secret key is a random permutation $(\sum_{i=1}^{u} n_i \times \sum_{i=1}^{u} n_i)$-matrix P. The public key matrix $\tilde{G}$ is the result of multiplying the concatenation of secret generator matrices and the matrix P: $\tilde{G} = [G_{C_1}|...|G_{C_u}]P$. The security of these cryptosystems is based on the assumption that for $u \geqslant 2$ it is computationally difficult to find in $\tilde{G}$ such $(k \times n_i)$-submatrices composed of n_i columns of $\tilde{G}$, which would be generator matrices of codes permutably equivalent to the codes C_i, $i = 1, ..., u$. In the present paper, we construct an algorithm for the efficient search for such submatrices in the case when several conditions are satisfied. One of the conditions is the decomposability of the square of the connected codes C_1,...,C_u into the direct sum of the squares of the codes C_i, $i \in \{1, ..., u\}$. An experimental assessment of the probability of fulfilling this condition for some Reed–Solomon codes, binary Reed–Muller codes, and Goppa code are also provided.

Keywords: Sidel'nikov-type systems · Cryptanalysis · Decomposability of codes

1 Introduction

Code-based cryptosystems are considered as a possible alternative to asymmetric cryptosystems, the strength of those is based on the complexity of the factorization problem or the discrete logarithm problem in the cyclic group [1]. In [2] R. McEliece proposed an asymmetric code cryptosystem based on the hardness of decoding a random linear code. For the cryptosystem described in [2], which is based on Goppa codes, an effective algorithm for finding a secret key by public key has not been found yet. However, since the public keys of Goppa code–based cryptosystem is very large, it makes sense to consider other codes. In general, the public key matrix $\tilde{G}$ of the McEliece-type system is of the form

$$\tilde{G} = SG_C P, \tag{1}$$

M. Baldi et al. (Eds.): CBCrypto 2020, LNCS 12087, pp. 20–40, 2020.
https://doi.org/10.1007/978-3-030-54074-6_2

where the nonsingular $(k \times k)$-matrix S, the permutation $(n \times n)$-matrix P and generator matrix G_C of $[n, k, d]_q$-code C form the secret key of the cryptosystem. Note that the code C should have a polynomial decoder. For fast decipherment, without knowing (S, P, C), it is sufficient to find a nonsingular $(k \times k)$-matrix S', a permutation matrix P' and a generator matrix $G_{C'}$ for linear $[n, k, d']$-code C' with a polynomial decoder that

$$S'G_{C'}P' = \tilde{G}$$

and $d' \geqslant d$. These attacks are called structural attacks or key attacks. The results of [3]–[7] show that this complex problem can be solved in polynomial time for some well-known codes. Unfortunately, an algorithm for finding a suitable key $(S', P', G_{C'})$ in the general case is unknown. By $\mathcal{S}_n$ we denote the symmetric group of degree n. Note that if adversary knows the code C, then for quick decipherment it is enough to solve the problem of finding a permutation $\sigma \in \mathcal{S}_n$ such that $\sigma(C) = \tilde{C}$ [8], where $\tilde{C}$ is a code generated by the rows of $\tilde{G}$. (Hereinafter by $\sigma(C)$ we mean the code obtained from C by permutation the coordinates in the codewords under σ.)

In [9] a new asymmetric code cryptosystem was proposed by V. M. Sidel'nikov. The public key of this cryptosystem is obtained from (1) by replacing SG_C with a concatenation of $u(\in \mathbb{N})$ randomly selected generator matrices for *one* fixed $[n, k, d]_q$-code C. The security of this cryptosystem against structural attacks is based on the assumption that for $u \geqslant 2$ in the public key matrix $\tilde{G}$ it is computationally hard to find u such $(k \times n)$-submatrices, composed of different n columns, where each submatrix generates some code permutably equivalent to the code C. In particular it was believed that for $u \geqslant 4$ such cryptosystems can be more secure than McEliece-type systems on Reed–Muller codes C_i, $i \in \{1, ..., u\}$ [9]. Note that for $u = 1$ the results of cryptanalysis Sidel'nikov-type system based on Reed–Muller codes are given in [6, 7], and for $u = 2$ such results are considered in [10]. In [11, 12] Sidel'nikov-type system is generalized by using concatenation of $u(\in \mathbb{N})$ randomly selected generator matrices for *different* codes C_1, ..., C_u of the same dimension.

In this paper, it is shown that for any u the problem of finding secret keys for cryptosystems from the [9, 11, 12] can be reduced to cryptanalysis of the McEliece's systems if some conditions are met. This paper is organized as follows. Section 2 provides the necessary information about linear codes and merged codes. Section 3 provides an algorithm from [13] for decomposing a linear code into a direct sum of indecomposable subcodes. In Sect. 4, using this, we construct a new algorithm for splitting the support of merged codes. This algorithm is used in Sect. 5 to attack the public key of Sidel'nikov-type cryptosystem. Also Sect. 5 presents the experimental results to assess the probability of satisfying conditions, sufficient for cryptanalysis for Sidel'nikov-type systems on some well-known codes. In particular, these results show that for $u \geqslant 4$ Sidel'nikov-type system on Reed–Muller codes can be broken by using [6, 7]. Note that for $u = 2$ the results of cryptanalysis Sidel'nikov-type system on some codes are given in [14].

2 Preliminaries

2.1 Linear Codes and Schur Product

By $\underline{n}$ denote the set $\{1, ..., n\}$. Let $\mathbf{x} = (x_1, ..., x_n) \in \mathbb{F}_q^n$; recall that the support of $\mathbf{x}$ is $\text{supp}(\mathbf{x}) = \{i \in \underline{n} : x_i \neq 0\}$. Cardinality of this set is called Hamming weight of $\mathbf{x}$, we denote it by $\text{wt}(\mathbf{x})$. The support of a set of vectors is the union of the supports of vectors from this set. In the space $\mathbb{F}_q^n$ we consider a linear $[n, k, d]_q$-code C of dimension k, effective length n and code distance d. By the effective length of code, we mean the cardinality of the set $\cup_{\mathbf{c} \in C} \text{supp}(\mathbf{c})$. Sometimes, when the code distance d is unknown or when its value is not important, we say that a $[n, k, d]_q$-code C is just $[n, k]_q$-code. By Dec_C we denote a decoding algorithm for C that can correct up to $t = \lfloor (d-1)/2 \rfloor$ errors. The output of Dec_C is an information vector from $\mathbb{F}_q^k$. Such decoder have to know the generator matrix G_C used in the encoding. Note that reconstructing the information vector from the code vector and G_C is a trivial procedure. Because of this, decoders are often defined as algorithms that find error vectors by the received ones.

It is said that a $[n, k, d]_q$-code C is permutably equivalent to a $[n, k, d]_q$-code D if there exists a permutation σ from the symmetric group $\mathcal{S}_n$ such that

$$D = \sigma(C) = \{\sigma(\mathbf{c}) = (c_{\sigma(1)}, ..., c_{\sigma(n)}) : \mathbf{c} = (c_1, ..., c_n) \in C\}.$$

For these codes we use the botation $C \sim D$. The group

$$\text{PAut}(C) = \{\sigma \in \mathcal{S}_n : \sigma(C) = C\}$$

is said to be the group of permutation automorphisms of C. Let P_σ be the permutation $(n \times n)$-matrix corresponding to $\sigma(\in \mathcal{S}_n)$. For $\mathbf{a} = (a_1, ..., a_n)$ and $\mathbf{b} = (b_1, ..., b_n)$ from $\mathbb{F}_q^n$ we consider component-wise multiplication

$$\mathbf{a} \star \mathbf{b} = (a_1 b_1, ..., a_n b_n);$$

it is also called the Schur product [15]. It is easy to verify that for any $\mathbf{a}, \mathbf{b}, \mathbf{c} \in \mathbb{F}_q^n$, $\sigma(\in \mathcal{S}_n)$, and $\alpha, \beta, \gamma \in \mathbb{F}_q$ the following equalities hold:

$$(\alpha \mathbf{a} + \beta \mathbf{b}) \star \gamma \mathbf{c} = \alpha\gamma(\mathbf{a} \star \mathbf{c}) + \beta\gamma(\mathbf{b} \star \mathbf{c}), \tag{2}$$

$$\sigma(\mathbf{a}) \star \sigma(\mathbf{b}) = \sigma(\mathbf{a} \star \mathbf{b}). \tag{3}$$

The square of the linear $[n, k]_q$-code C is the code generated by the set $\{\mathbf{a} \star \mathbf{b} : \mathbf{a}, \mathbf{b} \in C\}$; we denote it by C^2. If $\mathbf{g}_1$, ..., $\mathbf{g}_k$ are the rows of G_C then C^2 is generated by the rows of the $(k(k+1)/2 \times n)$-matrix $(\mathbf{g}_i \star \mathbf{g}_j)_{i=1, j \geqslant i}^k$. It is easy to show that for $[n, k]_q$-code C the following inequality holds

$$\dim(C^2) \leqslant \min\left\{n, \frac{k(k+1)}{2}\right\}. \tag{4}$$

In [16] for a random linear code C theoretical estimates on the probability of equality in (4) are obtained. For some well-known codes, exact formulas for the dimension of its squares have been found. Let $\mathrm{GRS}_{k,n}$ be a generalized $[n,k]_q$-Reed–Solomon code, $\mathrm{RM}(r,m)$ be a binary Reed–Muller code of order r and lengths 2^m. Using (2) one can show that

$$\mathrm{GRS}^2_{k,n} = \mathrm{GRS}_{\min\{2k-1,n\},n},$$

$$\mathrm{RM}(r,m)^2 = \mathrm{RM}(\min\{2r,m\},m)$$

(see [5,7,17]).

We need the following simple lemma.

Lemma 1. *Let D, K be $[n,k]_q$-codes, $\pi \in \mathcal{S}_n$. If $K = \pi(D)$, then*

$$K^2 = \pi(D^2). \tag{5}$$

Lemma 1 implies that $\pi(D^2) = (\pi(D))^2$, and if $(\pi(D))^2 = D^2$, then $\pi \in \mathrm{PAut}(D^2)$.

Using the definition of the square-code, follows that for any $D(\subseteq C)$ the embedding $D^2 \subseteq C^2$ holds. Finding conditions on C and D for $D^2 = C^2$ is of interest. Note that the square of subcode of GRS is more likely to be the GRS code of higher dimension. This fact is used in [5] to reduce the attack on the Berger-Loidreau cryptosystem to the cryptanalytic algorithm from [3]. In [18] a similar approach is applied to attack a key of the Berger-Loidreau system, which is based on Reed–Muller binary subcodes. In [19] finding the Reed–Solomon code by the square of the subcode is used to attack one modification of the McEliece-type cryptosystem, different from the Berger-Loidreau modification. In the following sections, we consider the case when a code D is a merging of u codes and a code C is a direct sum of these u codes.

2.2 The Construction of Merging of Linear Codes

Let $\mathcal{G}(C)$ be the set of all generator matrices of the code C, $\mathcal{L}(A)$ be the linear span of the rows of the matrix A. Let $\tau \subseteq \underline{n}$; by A_τ we denote the matrix composed of columns of the matrix A with indices from τ. For $[n_i,k,d_i]_q$-codes C_i, $i \in \underline{u}$, and $n = \sum_{i=1}^{u} n_i$ let's consider the family of $[n,k]_q$-codes

$$\mathcal{E}(C_1,...,C_u) = \{\mathcal{L}([G_{C_1}|...|G_{C_u}]) : G_{C_i} \in \mathcal{G}(C_i), i \in \underline{u}\}, \tag{6}$$

where $[A|B]$ is concatenation of A and B. Codes from $\mathcal{E}(C_1,...,C_u)$ will be called as *merged* codes (sometimes in literature, such codes are called as *connection of codes* [20]). Let $G_D = [G'_{C_1}|...|G'_{C_u}]$ be the generator matrix of the code D from $\mathcal{E}(C_1,...,C_u)$, let Dec_{C_i} be a decoder for C_i in which the *fixed* generator matrix $\hat{G}_{C_i}$ is used. By matrix G_D one can easily find (with complexity $\mathcal{O}(\sum_{i=1}^{u} n_i^3)$) such u nonsingular $(k_i \times k_i)$-matrices M_i that

$$G_D = [M_1\hat{G}_{C_1}|...|M_u\hat{G}_{C_u}]. \tag{7}$$

Each code D from this family has a code distance d_D of at least

$$d(C_1, ..., C_u) := \sum_{i=1}^{u} d_i,$$

i.e. $d_D \geqslant d(C_1, ..., C_u)$. Note that there is a code in the family whose minimum code distance is $d(C_1, ..., C_u)$. Hence the exhaustive decoder can correct up to

$$\lfloor (d_D - 1)/2 \rfloor \geqslant \lfloor (d(C_1, ..., C_u) - 1)/2 \rfloor$$

errors. In the general case, the problem of constructing a fast decoder for an arbitrary code $D(\in \mathcal{E}(C_1, ..., C_u))$ that corrects up to $\lfloor (d_D - 1)/2 \rfloor$ errors is not solved. If the vector has no more than $\lfloor (d(C_1, ..., C_u) - 1)/2 \rfloor$ errors, then the following method can be used. The method is based on the scheme from [9]; where a decoder for merged Reed–Muller binary codes is proposed. Let

$$\mathbf{m}[G'_{C_1}|...|G'_{C_u}] + \mathbf{e} = \mathbf{c}$$

be noisy codeword of D,

$$\text{wt}(\mathbf{e}) \leqslant t = \lfloor (d(C_1, ..., C_u) - 1)/2 \rfloor. \tag{8}$$

The vectors $\mathbf{c}$ and $\mathbf{e}$ can be represented as the merging of u subvectors:

$$\mathbf{c} = (\mathbf{c}_1, ..., \mathbf{c}_u), \mathbf{e} = (\mathbf{e}_1, ..., \mathbf{e}_u), \mathbf{c}_i, \mathbf{e}_i \in \mathbb{F}_q^{n_i}.$$

To decode $\mathbf{c}$, it is sufficient to find

$$\mathbf{m}'_i = \mathsf{Dec}_{C_i}(\mathbf{c}_i) M_i^{-1}, i \in \underline{u},$$

and choose the index i such that the inequality holds:

$$\text{wt}(\mathbf{c} - \mathbf{m}'_i G_D) \leqslant t. \tag{9}$$

In fact, the condition (8) implies that there is at least one such vector $\mathbf{e}_{j_0} (\in \mathbb{F}_q^{n_j})$ that $\text{wt}(\mathbf{e}_{j_0}) \leqslant t_{j_0} = \lfloor (d_{j_0} - 1)/2 \rfloor$. Therefore, among the vectors $\mathbf{m}'_1$, ..., $\mathbf{m}'_u$ there is at least one for such that (9) holds. In the general case, this inequality can hold for different $\mathbf{m}'_{i_0}$, $\mathbf{m}'_{j_0}$, $i_0 \neq j_0$. Hence, under the condition (8), decoding by the legal recipient may be wrong (this follows from the fact that the recipient, without knowing additional information about $\mathbf{m}'_{i_0}$ and $\mathbf{m}'_{j_0}$, cannot give preference to any information vector). Using a checksum in information vectors will make it possible to choose the message for which such sum would be correct. However, this method decreases the information coding rate. Another way to exclude decoding indetermination is to decrease the maximum number of correctable errors in a channel. For example, in [9] for merged Reed–Muller binary codes the maximum number of errors is $\sum_{i=1}^{u} t_i + u - 1$, $t_i = \lfloor (d_i - 1)/2 \rfloor$, and in [11] for two merged different codes the maximum number of errors is $t_1 + t_2$. Moreover, the correct decoding is always guaranteed in [9], and the probability of the correct decoding in [11] is close to 1.

Further, by $t_{\text{lim}}(\leqslant \lfloor (d(C_1, ..., C_u) - 1)/2 \rfloor)$ we mean the permissible number of errors that can be corrected in this way. Using the above consideration, it follows that for some families of codes, t_{lim} can be determined theoretically. For some other codes, the threshold can be estimated experimentally. The described decoding method will be denoted by $\mathsf{Dec}_{\mathcal{E}(C_1,...,C_u),t_{\text{lim}}}$.

3 Decomposition of Linear Codes

For linear $[n_i, k_i, d_i]_q$-codes C_i, $i \in \underline{u}$, and $n = \sum_{i=1}^{u} n_i$ a code

$$\mathcal{C} = C_1 \oplus ... \oplus C_u = \{(\mathbf{c}_1, ..., \mathbf{c}_u) : \mathbf{c}_i \in C_i\} \subseteq \mathbb{F}_q^n, \tag{10}$$

will be called as the external direct sum of u codes. Let $\tilde{C}_i(\subseteq \mathbb{F}_q^n)$ be a code obtained from the code $C_i(\subseteq \mathbb{F}_q^{n_i})$ by adding $n - n_i$ zero coordinates in the appropriate places. Then the code $\mathcal{C}$ is represented as inner direct sum

$$\mathcal{C} = \tilde{C}_1 + ... + \tilde{C}_u \subseteq \mathbb{F}_q^n, \ \text{supp}(\tilde{C}_i) \cap \text{supp}(\tilde{C}_j) = \varnothing. \tag{11}$$

Note that the representations (10) and (11) are equivalent, therefore, we will often write direct sum in short.

It is said that the code C is *decomposable* if this code is permutably equivalent to the direct sum of two or more nontrivial codes [21]. The code C is called *a decomposable code with decomposition length u* if it can be represented as the direct sum of u codes by the permutation of coordinates (see (10)) [13]. If in the decomposition (10) all codes C_i are indecomposable, then such decomposition will be called *complete.* Note that in this case the representation (10) is unique up to a permutation of the codes C_i in the sum. For convenience the indecomposable code C can be called as decomposable code with decomposition length one. Note that if the code C_i has the decomposition length v_i, $i \in \underline{u}$, then the code $C_1 \oplus ... \oplus C_u$ is a decomposable code with the decomposition length $v_1 + ... + v_u$. Thus, the decomposable code is permutably equivalent to the code with the block-diagonal generator matrix. Otherwise, the code is called *indecomposable.* Any maximum distance separable code (MDS) is an example of an indecomposable code [22].

An algorithm for the code decomposition (see Algorithm 1) was constructed in [13]. Recall that $\mathbf{c}(\in C)$ is minimal if there is no vector $\mathbf{c}'$ linearly independent with $\mathbf{c}$ such that $\text{supp}(\mathbf{c}') \subseteq \text{supp}(\mathbf{c})$ [23]. In [24] it is shown that for $[n,k]_q$-code C any generator matrix in a systematic form consists of minimal vectors of this code. The algorithm that brings the matrix to a systematic form is denoted by $\mathsf{Systematic}$. A set of vectors $M = \{\mathbf{x}_1, ..., \mathbf{x}_s\} \subseteq \mathbb{F}_q^n$ is called *connected* if there is such order $\mathbf{x}_{i_1}, ..., \mathbf{x}_{i_s}$ that for any $2 \leqslant k \leqslant s$

$$\text{supp}(\mathbf{x}_{i_k}) \cap \left(\cup_{j=1}^{k-1} \text{supp}(\mathbf{x}_{i_j})\right) \neq \varnothing.$$

All minimal vectors of the code C denote by $\mathcal{M}(C)$, the set of maximal connectivity components of C denote by $\mathcal{Z}(C)$ and any basis of minimal vectors is denoted by $\mathcal{B}_M(C)$.

Lemma 2. *Let C be a decomposable code of view* (11) *where $\tilde{C}_j$ is indecomposable subcode for all $i = 1, ..., u$. Then*

1) $\mathcal{M}(C) = \cup_{i=1}^{u} \mathcal{M}(\tilde{C}_i)$;
2) $\mathcal{Z}(C) = \{\mathcal{M}(\tilde{C}_1), ..., \mathcal{M}(\tilde{C}_u)\}$;
3) $\mathcal{B}_M(C) = \cup_{i=1}^{u} \mathcal{B}_M(\tilde{C}_i)$.

Proof. The first two statements follow directly from the conditions of the lemma and the definitions. The third statement follows from the first two.

Theorem 1. *Let C be a decomposable code with the complete decomposition length u. Then the algorithm* Decomposition *by an arbitrary generator matrix finds bases of indecomposable subcodes $\tilde{C}_1$, ..., $\tilde{C}_u$ whose supports do not intersect in pairs. The found bases consist of minimal code vectors and the complexity of the algorithm is $O(n^3 + k^2 n)$.*

Proof. At the first step of the algorithm Decomposition (see Algorithm 1), a basis $\mathcal{B}_M(C)$ is constructed from the minimal code vectors. The complexity of this step is $O(n^3)$. On the next steps the connected sets $\mathcal{B}_i$, $i = 1, ..., u$ are constructed from the found minimal basic code vectors. The supports of the sets $\mathcal{B}_i$ do not intersect and the basis $\mathcal{B}_M(C)$ constructed at the first step consists of minimal code vectors. Then it follows from item 3 of the lemma 2 that $\mathcal{B}_i = \mathcal{B}_M(\tilde{C}_i)$, and consists of minimal code vectors. The complexity of constructing the sets $\mathcal{B}_i$ is $O(k^2 n)$.

Algorithm 1. Decomposition

Input: G_C is generator matrix of $[n, k]_q$-code C
Output: $\mathcal{B}_1, ..., \mathcal{B}_u$ are bases of codes $\tilde{C}_i (\subseteq C)$, consisted of minimal vectors, such that $C = \tilde{C}_1 + ... + \tilde{C}_u$, $\text{supp}(\tilde{C}_i) \cap \text{supp}(\tilde{C}_j) = \varnothing$ for $i \neq j$

```
{g'_1, ..., g'_k} = Systematic(G_C), V = k, i = 1
while V ≠ ∅ do
   For any j ∈ V: V_i = {j}, W_i = supp(g'_j), changed = true
   while changed = true do
      changed = false
      for all l ∈ V \ {j} do
         if supp(g'_l) ∩ W_i ≠ ∅ then
            changed = true, V_i = V_i ∪ {l}, W_i = W_i ∪ supp(g'_l)
         end if
      end for
   end while
   B_i = {g'_j : j ∈ V_i}, V = V \ V_i, i = i + 1
end while
Return B_1, ..., B_u
```

The Decomposition algorithm is used in [13] for cryptanalysis the McEliece-type cryptosystem based on a direct sum of codes. The algorithm Decomposition can be modified to the algorithm DiagonalDecomposition with the same complexity. The last algorithm for the code C with complete decomposition (10) uses the generator matrix G_C in the systematic form and constructs such permutations $\sigma_l(\in \mathcal{S}_k)$, $\sigma_r(\in \mathcal{S}_n)$ that the matrix $P_{\sigma_l} G_C P_{\sigma_r}$ has a block-diagonal form:

$$P_{\sigma_l} G_C P_{\sigma_r} = \text{diag}(B_1, ..., B_u),$$

where B_i is the generator matrix of a code that is permutably equivalent to the code $C_{\rho(i)}$ for some $\rho \in \mathcal{S}_u$.

In [25] it was proved that for any code D from the family $\mathcal{E}(C, C)$, considered in the Sect. 2.2, the embedding $D^2 \subseteq C^2 \oplus C^2$ holds. The following lemma is a natural generalization of this result.

Lemma 3. *For any* $D \in \mathcal{E}(C_1, ..., C_u)$ *the relation holds:*

$$D^2 \subseteq C_1^2 \oplus ... \oplus C_u^2. \tag{12}$$

Further the rank of the matrix A will be denoted by $r(A)$. If the generator matrices G_D, G_{C_1}, ..., G_{C_u} are known then the checking whether the code D^2 is decomposable into direct sum of codes C_1^2, ..., C_u^2, i.e. whether equality holds in (12), can be performed using the sequential exclusion algorithm. To do this it is sufficient to calculate $r(G_D^2)$ and compare it with the sum of $r(G_{C_1}^2)$, ..., $r(G_{C_u}^2)$.

4 Splitting Algorithm for the Merged Codes

Let us consider the family $\mathcal{E}(C_1, ..., C_u)$ generated by $[n_i, k]_q$-codes C_i, $i \in \underline{u}$ (see (6)). For a code $D(\in \mathcal{E}(C_1, ..., C_u))$ and $n = \sum_{i=1}^{u} n_i$ let us define such group of all permutations $\Gamma(D)$ that

$$\sigma(D) \in \mathcal{E}(C_1, ..., C_u)$$

for σ from $\Gamma(D)$.

Lemma 4. *If* C_1,...,C_u *are indecomposable and* $D(\in \mathcal{E}(C_1, ..., C_u))$ *is such that* $D^2 = C_1^2 \oplus ... \oplus C_u^2$, *then*

$$\text{PAut}(C_1 \oplus ... \oplus C_u) \subseteq \Gamma(D) \subseteq \text{PAut}(C_1^2 \oplus ... \oplus C_u^2).$$

Proof. Left inclusion follows from the structure of PAut$(C_1 \oplus ... \oplus C_u)$ for indecomposable codes (see the paper by D. Slepian [21], Theorem 2).

Let $\sigma \in \Gamma(D)$, then there are nonsingular $(k \times k)$-matrices M_1 ,..., M_u that

$$\begin{aligned}([G_{C_1}|...|G_{C_u}]P_\sigma)^2 &= ([M_1 G_{C_1}|...|M_u G_{C_u}])^2 \\ &= ([M_1|...|M_u]\text{diag}(G_{C_1}, ..., G_{C_u}))^2\end{aligned}$$

for fixed generator matrices G_{C_1},...,G_{C_u}. So

$$\mathcal{L}(([M_1|...|M_u]\text{diag}(G_{C_1}, ..., G_{C_u}))^2) \subseteq C_1^2 \oplus ... \oplus C_u^2.$$

From (3) we obtain

$$([G_{C_1}|...|G_{C_u}]P_\sigma)^2 = ([G_{C_1}|...|G_{C_u}])^2 P_\sigma.$$

The condition of the theorem imply that

$$\text{rank}(([G_{C_1}|...|G_{C_u}])^2) = \sum_{i=1}^{u} \dim(C_i^2).$$

As a permutation of the columns does not change the rank of a matrix, then

$$\mathcal{L}(([M_1|...|M_u]\text{diag}(G_{C_1}, ..., G_{C_u}))^2) = C_1^2 \oplus ... \oplus C_u^2.$$

It means that $\sigma \in \text{PAut}(C_1^2 \oplus ... \oplus C_u^2)$. So right inclusion is proved.

Let K be a code permutably equivalent to some $[n,k,d]_q$-code from the family $\mathcal{E}(C_1, ..., C_u)$, G_K is any generator matrix of K. Consider the problem of finding the permutation $\pi(\in \mathcal{S}_n)$ such that $G_K P_\pi = [W_1|...|W_u]$ and $\mathcal{L}(W_i) \sim C_i$ for $i \in \underline{u}$.

Theorem 2. *Let C_i be a linear $[n_i, k]_q$ code, $i \in \underline{u}$, $n = \sum_{i=1}^{u} n_i$, and K be a $[n,k]_q$-code permutably equivalent to some unknown $[n,k]_q$-code D from the family $\mathcal{E}(C_1, ..., C_u)$ for which*

$$\Gamma(D) = \text{PAut}(C_1^2 \oplus ... \oplus C_u^2), \tag{13}$$

$$D^2 = C_1^2 \oplus ... \oplus C_u^2. \tag{14}$$

Suppose that one of two conditions is satisfied:

1) $C_1^2 = ... = C_u^2 = C$, and there are not subcodes with the same dimension and length in the complete decomposition of C,
2) for $i \neq j$ in the complete decomposition of the codes C_i^2 and C_j^2 there are no codes that coincide in dimension and effective length.

Then there exists a polynomial algorithm CodeSplitting*, which by an arbitrary generator matrix G_K of the code K finds such a permutation $\pi(\in \mathcal{S}_n)$ that*

$$G_K P_\pi = [W_1|...|W_u], \mathcal{L}(W_i) \sim C_i, i \in \underline{u}. \tag{15}$$

Proof. Let $\hat{C}_{i,1} \oplus ... \oplus \hat{C}_{i,R_i}$ be a complete decomposition of the code C_i^2, $i \in \underline{u}$, i.e.

$$\hat{C}_{i,1} \oplus ... \oplus \hat{C}_{i,R_i} \sim C_i^2.$$

From (14) it follows that the length of complete decomposition of D^2 is equal to

$$v = \sum_{i=1}^{u} R_i. \tag{16}$$

Note that if we apply the algorithm DiagonalDecomposition to the generator matrix G_{D^2} in the systematic form then we obtain the permutations $\phi_l(\in \mathcal{S}_{\hat{k}})$, $\phi_r(\in \mathcal{S}_n)$, such that

$$P_{\phi_l} G_{D^2} P_{\phi_r} = \text{diag}(..., A_{i,1}, ..., A_{i,R_i}, ...), \tag{17}$$

where $\hat{k} = \dim(D^2)$ and $\mathcal{L}(A_{i,l}) \sim \hat{C}_{i,s}$, $i \in \underline{u}$, $l, s \in \{1, ..., R_i\}$. By condition of the theorem, the codes C_1, ..., C_u are known, and the code D is unknown. However, the equality (14) holds for it. Therefore in (17) the matrix G_{D^2} can be selected in the form $\text{diag}(G_{C_1^2}, ..., G_{C_u^2})$, where $G_{C_i^2}$ is the generator matrix of C_i^2 in the systematic form, $i \in \underline{u}$.

By the condition of the theorem the codes K and K^2 are known, and codes D and K are permutably equivalent. Therefore, by the lemma 1 the codes D^2 and K^2 are also permutably equivalent. Let G_{K^2} be an arbitrary generator matrix of the code K^2 in the systematic form. We can use the algorithm DiagonalDecomposition for this matrix to find a permutations $\sigma_l (\in \mathcal{S}_{\hat{k}})$ and $\sigma_r (\in \mathcal{S}_n)$ such that

$$P_{\sigma_l} G_{K^2} P_{\sigma_r} = \text{diag}(B_1, ..., B_v). \tag{18}$$

Suppose that there is an effective algorithm ArrangeBlocks, which sequentially for each $i \in \underline{u}$ selects R_i blocks B_{j_1}, ..., $B_{j_{R_i}}$ in the matrix (18) so that

$$C_i^2 \sim \mathcal{L}(\text{diag}(A_{i,1}, ..., A_{i,R_i})) \sim \mathcal{L}(\text{diag}(B_{j_1}, ..., B_{j_{R_i}})) = \hat{D}_i.$$

A method for constructing ArrangeBlocks will be described below under condition 1) or 2) from the theorem. In the theorem it is supposed that $K = \pi'(D)$ and π' is unknown. Thus, using ArrangeBlocks one can find permutations $\delta_l (\in \mathcal{S}_{\hat{k}})$, $\delta_r (\in \mathcal{S}_n)$ such that

$$P_{\delta_l} P_{\sigma_l} G_{K^2} P_{\sigma_r} P_{\delta_r} = P_{\delta_l} P_{\sigma_l} G'_{D^2} P_{\pi'} P_{\sigma_r} P_{\delta_r} = \text{diag}(B'_1, ..., B'_u), \tag{19}$$

where G'_{D^2} is some generator matrix of the code D^2 in the systematic form, $B'_i = G_{\hat{D}_i}$ is the generator matrix of the code $\hat{D}_i$, permutably equivalent to the code C_i^2, $i \in \underline{u}$. The permutation matrix $P_{\pi'} P_{\sigma_r} P_{\delta_r}$ can be represented as product $P_\alpha \text{diag}(P_{\beta_1}, ..., P_{\beta_u})$, where $\alpha \in \text{PAut}(C_1^2 \oplus ... \oplus C_u^2)$, $\beta_i \in \mathcal{S}_{n_i}$, $i \in \underline{u}$. Since (13), then $\alpha \in \Gamma(D)$. Therefore, for some (unknown) generator matrix G'_D of code D the matrix $G_K P_{\sigma_r} P_{\delta_r}$ can be represented in the form

$$G_K P_{\sigma_r} P_{\delta_r} = G'_D P_{\pi'} P_{\sigma_r} P_{\delta_r} = G'_D P_\alpha \text{diag}(P_{\beta_1}, ..., P_{\beta_u}). \tag{20}$$

As $\alpha \in \Gamma(D)$ then there is a code D' in $\mathcal{E}(C_1, ..., C_u)$ that

$$G_K P_{\sigma_r} P_{\delta_r} = G_{D'} \text{diag}(P_{\beta_1}, ..., P_{\beta_u}) = [W_1|...|W_u],$$

where $\mathcal{L}(W_i) \sim C_i$, $i \in \underline{u}$. So, we have

$$\pi = \delta_r \circ \sigma_r. \tag{21}$$

Now we show how under conditions 1) or 2) the algorithm ArrangeBlocks for finding permutations δ_r, δ_l can be constructed.

Let condition 1) be hold. Then (16) will take the form $v = uR$, where R is the length of complete decomposition of C. If C is an indecomposable code, then $R = 1$, $v = u$. In this case in the block-diagonal representations (17) and (18) there will be u blocks, all the same size (due to the permutative equivalence of the codes D^2 and K^2, as well as the uniqueness of the complete decomposition of these codes up to a permutation of the code summands). Therefore, in the algorithm ArrangeBlocks we can set δ_l and δ_r equal to the identity permutations. If C is a decomposable code, but in the complete decomposition there are no codes that match both the dimension and the effective length, then the set of all v blocks of the matrix (17) splits into R different classes. Each class contains u blocks of the same dimension and the same effective length. The same can be said about the blocks of the matrix (18). The bijective correspondence in the algorithm ArrangeBlocks between R classes for the matrix (17) and R classes for the matrix (18) is established naturally considering the size of the matrices. In the algorithm ArrangeBlocks within each class, a bijective correspondence between u blocks of (17) and u blocks of (18) is set arbitrarily. The choice of such a correspondence will only affect the final permutation π, which is not unique.

If condition 2) is fulfilled, then in the complete decomposition of $\phi_r(D^2)$ and $\sigma_r(K^2)$ there are no subcodes that coincide in both dimension and effective length and in the same time corresponding to different codes C_i^2 and C_j^2. Therefore, in the algorithm ArrangeBlocks the correspondence between the blocks in (17) and (18) is also established naturally taking into account the size of the matrices.

The algorithm CodeSplitting (see the Algorithm 2) implements the described above method for finding π. The input of this algorithm is the matrices G_K, G_{C_1}, ..., G_{C_u}, and the output is the permutation π. The complexity of finding (σ_l, σ_r) does not exceed the complexity of DiagonalDecomposition applied to the matrix G_{K^2}. Since this matrix has at most $k(k+1)/2$ rows and n columns, the complexity of the algorithm DiagonalDecomposition is $\mathcal{O}(n^3 + k^2 n))$. Under conditions 1) and 2) of the theorem, the permutations (δ_l, δ_r) can be found in polynomial time: the complexity does not exceed $\mathcal{O}(2kn)$.

Algorithm 2. CodeSplitting

Input: G_K
Output: π
  $G' = \mathsf{Systematic}(G_K^2)$
  $(\sigma_l, \sigma_r) = \mathsf{DiagonalDecomposition}(G')$
  $(\delta_l, \delta_r) = \mathsf{ArrangeBlocks}(P_{\sigma_l} G' P_{\sigma_r})$
Return $\pi = \delta_r \circ \sigma_r$

5 Structural Cryptanalysis of Sidel'nikov-Type Systems

5.1 Sidel'nikov-Type Cryptosystems

The code cryptosystem proposed by R. McEliece in [2] is based on the generator matrix G_C of $[n, k, d]_q$-code C, randomly selected nonsingular $(k \times k)$-matrix

S and permutation $(n \times n)$-matrix P [2]. The secret key is the triple (S, P, C), and the public one is the pair $(\tilde{G}, t)$, where the matrix $\tilde{G}$ has the form (1), $t = \lfloor (d-1)/2 \rfloor$. The encryption rule for the vector $\mathbf{m}(\in \mathbb{F}_q^k)$ has the form:

$$\mathbf{c} = \mathbf{m}\tilde{G} + \mathbf{e}, \ \mathrm{wt}(\mathbf{e}) \leqslant t, \tag{22}$$

where $\mathbf{e}$ is randomly selected error vector. The rule $\mathbf{m} = S^{-1}\mathsf{Dec}_C(\mathbf{c}P^{-1})$ is applied for decryption. A McEliece-type cryptosystem on the code C will be denoted by McE(C).

It was noted above that for a few linear codes a McEliece-type system is not resistant to attacks on a key (see [7]–[8]). For McE(C) let us denote by Attack_C an known algorithm of structural attack that finds the suitable secret key by the public key of McE(C). If the code C is known for adversary then the algorithm Attack_C can be based on the support splitting algorithm [8]. Note that the best known algorithm Attack_C may be non polynomial. Known polynomial attack algorithms include Sidel'nikov–Shestakov algorithm [3] and Wieschebrink [5] algorithm for generalized Reed–Solomon codes $\mathrm{GRS}_{k,n}$, the Minder–Shokrollahi algorithm [6] and the Borodin–Chizhov algorithm [7] for Reed–Muller binary codes $\mathrm{RM}(r, m)$.

In order to increase strength, some modifications of this system are proposed. In [9] V. M. Sidel'nikov proposed a system based on the merging of generator matrices of single code. Let G_C be the generator matrix of the $[n, k, d]_q$-code C, $u(\in \mathbb{N})$ be the system parameter. The public key $\tilde{G}$ matrix is

$$\tilde{G} = [G_1|...|G_u]P, \ \ G_i = M_i G_C \ i \in \underline{u}, \tag{23}$$

where M_i are randomly selected nonsingular $(k \times k)$-matrices, P is a randomly selected permutation $(un \times un)$-matrix. The encryption rule has the form (22), where $t = t_{\mathrm{lim}}$ is allowable number of errors that the decoder can correct for code from the family $\mathcal{E}(C_1, ..., C_u)$, with $C_i = C$, $i \in \underline{u}$. For decryption it is enough to decode the vector $\mathbf{c}P^{-1}$. For decoding one may use the decoder $\mathsf{Dec}_{\mathcal{E}(C_1,...,C_u),t_{\mathrm{lim}}}$ for $C_1 = ... = C_u = C$ (see the Sect. 2.2). This cryptosystem further is denoted by $\mathrm{Sid}_u(C)$.

The apparent generalization of the Sidel'nikov-type system is a cryptosystem based on merging $u(\in \mathbb{N})$ generator matrices of u *different* codes [11,12]. Let C_i be $[n_i, k, d_i]_q$-code with the generator matrix G_{C_i}, $i \in \underline{u}$, $n = \sum_{i=1}^{u} n_i$. Then the public key matrix $\tilde{G}$ has the form (23), where $G_i = M_i G_{C_i}$, $i \in \underline{u}$. Encryption and decryption is performed similarly to the corresponding operations in $\mathrm{Sid}_u(C)$. We will denote such a cryptosystem by SidMod$(C_1, ..., C_u)$. Note that code $\mathcal{L}(\tilde{G})$ is permutably equivalent to some code from $\mathcal{E}(C_1, ..., C_u)$.

5.2 Structural Attacks

One way to obtain information about the secret key of SidMod$(C_1, ..., C_u)$ by the corresponding public key $\tilde{G}$ of the form (23) is to obtain the matrix $\tilde{G}^2$ (using Schur product) and further attempt *to decompose* the code $K = \mathcal{L}(\tilde{G}^2)$

into the direct sum of subcodes. Indeed the Schur product was used earlier in the cryptanalysis of the Sidel'nikov-type system $\mathrm{Sid}_u(C)$ for $u = 2$, namely the structure of the set of equivalent keys for $\mathrm{Sid}_2(C)$ was studied in [17,25] for the binary Reed–Muller code and in [25] for the Reed–Solomon code. In [10] for binary Reed–Muller code C an algorithm for finding a suitable secret key by the public key of the system $\mathrm{Sid}_2(C)$ was constructed.

As $\mathrm{Sid}_u(C) = \mathrm{SidMod}(C_1, ..., C_u)$ for $C_1 = ... = C_u = C$ therefore further we will consider the cryptosystem $\mathrm{SidMod}(C_1, ..., C_u)$ and use CodeSplitting to reduce the cryptanalysis of this system to the cryptanalysis of $\mathrm{McE}(C_i)$ for $i \in \underline{u}$.

Theorem 3. *Let* $(\tilde{G}, t_{\mathrm{lim}})$ *be the public key of the system* $\mathrm{SidMod}(C_1, ..., C_u)$ *and conditions of the Theorem 2 for* $K = \mathcal{L}(\tilde{G})$ *and codes* C_1,...,C_u *be satisfied. Then the computational complexity of structural attack on the system* $\mathrm{SidMod}(C_1, ..., C_u)$ *does not exceed*

$$\mathcal{O}(n^3 + k^2 n + 2kn + \sum_{i=1}^{u} Q(C_i)), \tag{24}$$

where $Q(C_i)$ *is the complexity of the* Attack_{C_i}.

Proof. Since the conditions of Theorem 2 are satisfied, the algorithm CodeSplitting can be applied to the code K to find a permutation π such that (15) is fulfilled. Therefore the matrix $\tilde{G}P_\pi$ can be represented as concatenation of public keys for the systems $\mathrm{McE}(C_i)$:

$$\tilde{G}P_\pi = [W_1|...|W_u].$$

For $i \in \underline{u}$ one can apply attacks Attack_{C_i} to the corresponding matrix W_i for finding suitable secret key. To decrypt the vector $\mathbf{c}$ obtained by the rule (22) for $t = t_{\mathrm{lim}}$, it suffices to decode the vector $\pi^{-1}(\mathbf{c})$ using suitable keys in the decoding algorithm like $\mathsf{Dec}_{\mathcal{E}(C_1,...,C_u),t_{\mathrm{lim}}}$.

So, the complexity of the structural attack does not exceed (24).

Remark 1. Let the conditions of Theorem 3 be satisfied. Then the cryptanalysis of the system $\mathrm{SidMod}(C_1, ..., C_u)$ is reduced to the cryptanalysis of the systems $\mathrm{McE}(C_i)$, $i \in \underline{u}$. Moreover if for some $\mathrm{McE}(C_i)$ there are polynomial algorithms Attack_{C_i}, then there is a polynomial algorithm to decipher messages in some cases. Suppose, for example, such an algorithm is known only for the cryptosystem $\mathrm{McE}(C_1)$. Then the adversary, having the permutation π of the view (21), can correctly decipher the message if the first n_1 coordinates of the vector $\pi(\mathbf{e})$ have no more than $\lfloor (d_1 - 1)/2 \rfloor$ nonzero elements. So, given the random nature of errors in the rule (22), an adversary can effectively decipher a fraction of encrypted vectors. Moreover if $\mathsf{Attack}_{C_{i_1}}$,..., $\mathsf{Attack}_{C_{i_r}}$ are polynomial algorithms, then the size of such fraction increases and depends on d_{i_1}, ..., d_{i_r}.

Remark 2. Theorem 3 impies that the reducibility of the cryptanalysis of system $\mathrm{SidMod}(C_1, ..., C_u)$ to u attacks on public keys of McEliece-type systems on codes C_1, ..., C_u is possible in some cases. Since the fulfillment of equality (14) is one of the conditions of theorem, and since it is probabilistic, so, in the present paper we experimentally study its probability. By $p_R(C_1, ..., C_u)$ we denote the probability that the rank of the square of the matrix (7) is equal to R, when matrices M_1,...,M_u are selected uniformly random from $\mathrm{GL}_k(\mathbb{F}_q)$. Since column permutation does not change rank of matrix, it follows that $p_R(C_1, ..., C_u)$ is also probability that the rank of the square of the matrix (23) is equal to R. Note that $p_R(C_1, ..., C_u)$ for $R = \sum_{i\in\underline{u}} \dim(C_i^2)$ is the probability of equality (14) fulfillment for randomly selected D. In [14] for some cryptosystems of Sidel'nikov-type, the results of an experimental estimate of $p_R(C_1, ..., C_u)$ for $u = 2$ were given.

Even not all conditions of Theorem 3 are fulfilled, in some cases one can find a key that allows to decipher some (perhaps not all) messages.

For example even (14) does not hold it is possible to decipher some messages if one can find columns in an arbitrary order of at least one matrix G_i in the public matrix $\tilde{G}$ of the form (23). For this one can use Schur product for the public key $\tilde{G}$, try to decompose the code $\mathcal{L}(\tilde{G})^2$ into direct sum of codes and find comlumns of unknown matrix using connectivity components.

Without loss of generality, suppose that an adversary knows the column indices of the matrix $G_1 = M_1 G_{C_1}$ in the matrix (23) (the order of the columns does not matter). We denote the set of these indices by τ, $|\tau| = n_1$, and by A_τ we denote as usual the matrix composed of the columns of the matrix A with the indices from τ. Let $\mathbf{c}$ be the ciphertext of the system $\mathrm{SidMod}(C_1, ..., C_u)$ obtained by the (22) rule. It follows from this rule that

$$\mathbf{c}_\tau = \mathbf{m}\tilde{G}_\tau + \mathbf{e}_\tau.$$

Then the adversary can apply the well-known key attack algorithm for $\mathrm{McE}(C_1)$ to the matrix $\tilde{G}_\tau$ and find a suitable key for it. This key can be used further to decipher $\mathbf{c}_\tau$ into $\mathbf{m}'(\in \mathbb{F}_q^k)$. In the case

$$\mathrm{wt}(\mathbf{e}_\tau) \leqslant \lfloor (d_1 - 1)/2 \rfloor \tag{25}$$

the original clear text and decrypted vectors are the same: $\mathbf{m}' = \mathbf{m}$. An adversary can establish the correct decipherment by checking the inequality

$$\mathrm{wt}(\mathbf{m}'\tilde{G} - \mathbf{c}) \leqslant t.$$

If (25) does not hold, then the equality $\mathbf{m}' = \mathbf{m}$ is not guaranteed. Thus, given the random nature of $\mathbf{e}$, it is possible to decipher not all ciphertexts, but only those for which (25) is fulfilled. Obviously, the number of successful decipherments will be the greater, the more an adversary can find the corresponding columns of different G_i in (23).

In many cases the equality (14) holds (see the next section with experimental results) but it is not known whether the equality (13) holds. Thus, in general case we cannot directly apply found permutation π of the view (21) to transform the public key $\tilde{G}$ of Sidel'nikov-type system to concatenation of the public keys of McEliece-type systems. But if we additionaly know that for some code C_i^2

it is unlikely that such a code $C'(\dim(C') = k)$ exists that $(C')^2 = C_i^2$ and $C' \not\sim C_i$. Then we can try to find the columns in $\tilde{G}$ that form public key W_i for $\mathrm{McE}(C_i)$. As a result, one can apply algorithm Attack_{C_i} to W_i and perform partial decipherment with suitable secret key. For this case the corresponding example will be considered in the next subsection.

5.3 Examples and Experimental Results

Sid$_u$(GRS$_{k,n}$)-System. Recall that the GRS code $\mathrm{GRS}_{k,n} = \mathrm{GRS}_{k,n}(\mathbf{x}, \mathbf{y})$ is given by the vector $\mathbf{x} = (x_1, ..., x_n)$, where x_1, ..., x_n are pairwise distinct elements of the field $\mathbb{F}_q$, $n \leqslant q + 1$, and the vector $\mathbf{y} = (y_1, ..., y_n)$ of nonzero elements of the field $\mathbb{F}_q$. The generator matrix G_C for $C = \mathrm{GRS}_{k,n}(\mathbf{x}, \mathbf{y})$ has the form:

$$G_C = \begin{pmatrix} x_1^0 y_1 & \dots & x_n^0 y_n \\ x_1^1 y_1 & \dots & x_n^1 y_n \\ \vdots & \ddots & \vdots \\ x_1^{k-1} y_1 & \dots & x_n^{k-1} y_n \end{pmatrix}. \tag{26}$$

Equation (26) implies that $\mathrm{GRS}_{k,n}^2(\mathbf{x}, \mathbf{y}) = \mathrm{GRS}_{\min\{2k-1,n\},n}(\mathbf{x}, \mathbf{y} \star \mathbf{y})$. For the experiment, the parameters $q = n \in \{257, 523\}$, $k \in \{1, ..., \lfloor n/2 \rfloor\}$, $u \in \{2, ..., 8\}$ are chosen. For each triple (n, k, u) 100000 experiments were carried out. In each experiment the square of a matrix (7) is constructed and rank of the resulting matrix is calculated ($C_1 = ... = C_u = \mathrm{GRS}_{k,n}$). Note that for the chosen parameters k and n the equality $\min\{2k - 1, n\} = 2k - 1$ holds. Therefore, we have

$$\dim(C_1^2 \oplus ... \oplus C_u^2) = \sum_{i \in \underline{u}} r(G_{C_i}^2) = u(2k - 1). \tag{27}$$

Using (4) and (27), we obtain that $p_R(C_1, ..., C_u) = 0$ for all k such that

$$\frac{k^2 + k}{4k - 2} < u. \tag{28}$$

The experiments showed that for all considered triples (n, k, u) and for all k for which the inequality (28) holds, except for the triple $(257, 6, 2)$, the rank of the square of the matrix (7) is equal to $k(k + 1)/2$ in all experiments. In the case of $(n, k, u) = (257, 6, 2)$, the rank of such a matrix is $k(k + 1)/2 - 1 = 20$ in three experiments and in other cases it is equal to $k(k+1)/2 = 21$. Let (28) not hold, then the rank of the square of matrix (7) is $u(2k-1)$. Consider the triplets (n, k, u), let (28) not hold for k, then using (27), we see that the probability is $p_R(C_1, ..., C_u)$ is close to one for $R = u(2k - 1)$. In other words, for the most considered codes D from $\mathcal{E}(C_1, ..., C_u)$, we have $D^2 = C_1^2 \oplus ... \oplus C_u^2$. Let (28) be violated. Since MDS codes are indecomposable, using the obtained experimental results and Theorem 2 we see that for the considered parameters n, k and u, a structural attack on a Sidel'nikov-type cryptosystem based on GRS codes can be reduced in polynomial time to u attacks on the cryptosystem $\mathrm{McE}(\mathrm{GRS}_{k,n})$

in the case $\Gamma(D) = \mathrm{PAut}(C_1^2 \oplus ... \oplus C_u^2)$. Note that for the system $\mathrm{McE}(\mathrm{GRS}_{k,n})$ there are effective structural attacks for all n (see [3–5]). Thus, a Sidel'nikov-type cryptosystem based on GRS codes with the considered parameters can be effectively broken. It seems that the probability $p_{u(2k-1)}(C_1, ..., C_u)$ is close to one for the other (not considered) triples (n, k, u) such that (28) does not hold. The theoretical justification of this assumption is of interest.

$\mathbf{Sid}_u(\mathbf{RM}(r, m))$-System. Let us consider the case $C_1 = ... = C_u = \mathrm{RM}(r, m)$. Recall that the generator matrix of the code $\mathrm{RM}(r, m)$ consists of $r + 1$ submatrices and generally has the form

$$G_{\mathrm{RM}(r,m)} = \left(G_0^\top | G_1^\top | \cdots | G_r^\top\right)^\top ,$$

here G_0 is the (1×2^m)-matrix of units, G_1 is $(m \times 2^m)$-matrix such that a column with the number i (starting from zero) represents the binary notation of the number i; and G_i block for $1 < i \leqslant m$ consists of all possible Schur products of i rows from the matrix G_1. Since the matrix G_1 contains only m rows it follows that the matrix G_i for $1 < i \leqslant m$ contains only $\frac{m!}{i!(m-i)!}$ rows.

Lemma 5. *Let $r = 0, ..., m - 1$; then $\mathrm{RM}(r, m)$ is an indecomposable code.*

Proof. Note that the code $\mathrm{RM}(0, m)$ is a repetition code generated by a single unit vector. The indecomposability of this code is obvious. Since $\mathrm{RM}(m - 1, m)$ is dual code of $\mathrm{RM}(0, m)$ it follows that $\mathrm{RM}(m - 1, m)$ is indecomposable. The indecomposability of the codes $\mathrm{RM}(1, m)$ and $\mathrm{RM}(m - 2, m)$ was shown, e.g. in [13] (p. 94, Example 2). Let us show that the code $\mathrm{RM}(m - 3, m)$ is indecomposable. In [7] it was proved that

$$\mathrm{RM}(r_1, m) \star \mathrm{RM}(r_2, m) = \mathrm{RM}(\min\{r_1 + r_2, m\}, m). \tag{29}$$

Hence $\mathrm{RM}(m-2, m) = \mathrm{RM}(m-3, m) \star \mathrm{RM}(1, m)$. Since the codes $\mathrm{RM}(m-2, m)$ and $\mathrm{RM}(1, m)$ are indecomposable, it follows that the code $\mathrm{RM}(m - 3, m)$ is also indecomposable. Similarly, the indecomposability of the code $\mathrm{RM}(m - i, m)$ implies the indecomposability of the code $\mathrm{RM}(m - i - 1, m)$ for $i = 1, ..., m - 3$.

It is known (see [26], Chap. 13, §9) that

$$\mathrm{PAut}(\mathrm{RM}(1, m)) = ... = \mathrm{PAut}(\mathrm{RM}(m - 2, m)).$$

So, using lemma 5, equality (29) and result of D. Slepian about decomposition of linear codes (see [21], Theorem 2), we obtain

$$\mathrm{PAut}(C_1 \oplus ... \oplus C_u) = \mathrm{PAut}(C_1^2 \oplus ... \oplus C_u^2) \tag{30}$$

for $C_1 = ... = C_u = \mathrm{RM}(r, m)$ and $r = 1, ..., \lfloor (m - 2)/2 \rfloor$.

Lemma 6. *Let $C_1 = ... = C_u = \mathrm{RM}(1, m)$. Then $D^2 \neq C_1^2 \oplus ... \oplus C_u^2$ for any D from $\mathcal{E}(C_1, ..., C_u)$.*

Proof. It suffices to coinsider the case $u = 2$. Let $G_D = [G_1|G_2]$ be any generator matrix of D, $\mathcal{L}(G_1) = \mathcal{L}(G_2) = \mathrm{RM}(1,m)$. The definition of $\mathrm{RM}(1,m)$ implies that the matrices G_1 and G_2 have at least one pair of the same columns. So we conclude that the systematic form of G_D^2 contains a connectivity component of the length al least $2^m + 1$ (note that the code $\mathrm{RM}(1,m)$ is indecomposable). Then $D^2 \neq C_1^2 \oplus C_2^2$.

So, by Lemmas 4–6 and (30) we obtain the following theorem.

Theorem 4. *Let* $C_1 = ... = C_u = \mathrm{RM}(r,m)$ *and* $r = 2, ..., \lfloor (m-2)/2 \rfloor$. *If* (14) *is hold for D from* $\mathcal{E}(C_1, ..., C_u)$, *then* (13) *is hold, i.e.*

$$\Gamma(D) = \mathrm{PAut}(C_1^2 \oplus ... \oplus C_u^2).$$

Thus, all conditions of Theorem 2 are satisfied in the case (14) and one can use the algorithm CodeSplitting to find permutation π by the code $K = \pi'(D)$.

To get estimation of $p_R(C_1, ..., C_u)$ the codes $\mathrm{RM}(r,m)$ are considered for $r \in \{2,3\}$, $m \in \{5,6,7,8\}$, $u \in \{2, ..., 18\}$. By $R_{max}(r,m)$ we denote the maximum possible rank of a square of matrix of the form (23) in the case $C_i = \mathrm{RM}(r,m)$. Let also $R(r,m,u) = u \cdot \dim(\mathrm{RM}(2r,m))$ be the dimension of the direct sum of u squares of the code $\mathrm{RM}(r,m)$. For each triple (r,m,u), 100000 experiments were carried, in each case the probability $p_R(C_1, ..., C_u)$ was estimated. The experimental results are shown in the Table 1. Since the codes C_1, ..., C_u are the same and their squares are indecomposable then Theorem 4 implies that if (14) holds, then all conditions of Theorem 3 are fulfilled. Note that for a McEliece-type cryptosystem on the Reed–Muller binary codes, effective attacks are known (see [6,7]). Therefore, for the parameters considered above, a Sidel'nikov-type cryptosystem on these codes can also be broken with high probability, but in [9] it was supposed that these cryptosystems are highly resistant for $u \geqslant 4$. We also note that as m increases, the probability of (14) tends to 1.

SidMod-System on Some Binary Reed–Muller and Goppa Codes. Let's consider $\mathrm{SidMod}(\mathrm{RM}(4,10), \mathrm{Goppa}_{386,512})$-system based on the $[1024, 386]_2$-code $\mathrm{RM}(4,10)$ and the binary $[512, 386]_{2^{29}}$-Goppa code $\mathrm{Goppa}_{386,512}$ with a design distance of 29 [26] (a "toy example" from [12]). Let D be the $[1535, 386]_2$-code permutably equivalent to the code with the generator public key matrix $\tilde{G}$. In our work, computational experiments were carried out, showed that, with the probability close to 1, the equality in (12) holds. Note that $[1024, 1013]_2$-code $\mathrm{RM}(4,10)^2 = \mathrm{RM}(8,10)$ is indecomposable and the length of $[512, 386]_{2^{29}}$-code $\mathrm{Goppa}_{386,512}$ is less than the length of $\mathrm{RM}(4,10)$. Therefore, in the complete decomposition of D^2, a code that is permutably equivalent to $\mathrm{RM}(4,10)^2$ can be found easily. Note it is unknown whether the equality

$$\Gamma(D) = \mathrm{PAut}(\mathrm{RM}(8,10) \oplus (\mathrm{Goppa}_{386,512})^2)$$

holds in this case (see (13)). Nevertheless, for randomly generated $\pi' \in \mathcal{S}_{1536}$ after applying DiagonalDecomposition we get only one block with dimension $\dim(\mathrm{RM}(8,10))$ and length 1024 in complete decomposition of the code

Table 1. The results of estimating $p_R(C_1, ..., C_u)$, $C_i = \mathrm{RM}(r, m)$, $i \in \underline{u}$.

(a) $C_i = \mathrm{RM}(2, 5)$

		$(R; p_R)$
u	2	$(60; 8 \cdot 10^{-5})$, $(61; 0.01587)$, $(\mathbf{62} = R(2,5,2); \mathbf{0.98405})$
	3	$(90; 3 \cdot 10^{-5})$, $(91; 0.001)$, $(92; 0.04504)$, $(\mathbf{93} = R(2,5,3); \mathbf{0.95393})$
	4	$(120; 2 \cdot 10^{-5})$, $(121; 0.00012)$, $(122; 0.00418)$, $(123; 0.08681)$, $(\mathbf{124} = R(2,5,4); \mathbf{0.90887})$
	$\geqslant 5$	$(136 = R_{max}(2,5); 1)$

(b) $C_i = \mathrm{RM}(2, 6)$

		(R, p_R)
u	2	$(113; 0.0011)$, $(\mathbf{114} = R(2,6,2); \mathbf{0.9989})$
	3	$(170; 0.00305)$, $(\mathbf{171} = R(2,6,3); \mathbf{0.99695})$
	4	$(226; 1 \cdot 10^{-5})$, $(227; 0.00569)$, $(\mathbf{228} = R(2,6,4), \mathbf{0.9943})$
	$\geqslant 5$	$(253 = R_{max}(2,6); 1)$

(c) $C_i = \mathrm{RM}(2, 7)$

		(R, p_R)
u	2	$(197; 2 \cdot 10^{-5})$, $(\mathbf{198} = R(2,7,2); \mathbf{0.99998})$
	3	$(296; 8 \cdot 10^{-5})$, $(\mathbf{297} = R(2,7,3); \mathbf{0.99992})$
	4	$(395; 0.00023)$, $(\mathbf{396} = R(2,7,4); \mathbf{0.99977})$
	$\geqslant 5$	$(435 = R_{max}(2,7); 1)$

(d) $C_i = \mathrm{RM}(2, 8)$

		(R, p_R)
u	2-4	$(R(2,8,u); \mathbf{1})$
	$\geqslant 5$	$(703 = R_{max}(2,8); 1)$

(e) $C_i = \mathrm{RM}(3, 7)$

		(R, p_R)
u	2-16	$(R(3,7,u); \mathbf{1})$
	$\geqslant 17$	$(2080 = R_{max}(3,7); 1)$

(f) $C_i = \mathrm{RM}(3, 8)$

		(R, p_R)
u	2-17	$(R(3,8,u); \mathbf{1})$
	$\geqslant 18$	$(4371 = R_{max}(3,8); 1)$

$(\pi'(D))^2$. In our 10000 experiments we found that in systematic generator matrix of $(\pi'(D))^2$ the connectivity component with length 1024 and dimension $\dim(\mathrm{RM}(8, 10))$ always corresponds to the 1024 columns of $\tilde{G}$ which generate the code V and $V \sim \mathrm{RM}(4, 10)$. So, we conclude that, with high probability, using the algorithm CodeSplitting one can find such permutation π that

$$\tilde{G}P_\pi = [W_1|W_2],$$

where $\mathcal{L}(W_1) \sim \mathrm{McE}(\mathrm{RM}(4, 8))$. To find the secret permutation for $\mathrm{McE}(\mathrm{RM}(4, 8))$ by W_1, one can use algorithms from [6] or [7]. Currently there is no efficient algorithm $\mathsf{Attack}_{\mathsf{Goppa}}$ for Goppa's code. So, using the obtained partial secret key, it is possible to efficiently decipher some ciphertexts (see the Remark 1).

6 Conclusion

In [13] it is shown that the analysis of McEliece-type system on the direct sum of codes can be reduced to the analysis of McEliece-type systems on the summands. In this paper, we have shown that the analysis of Sidel'nikov-type systems can also be reduced to the analysis of McEliece-type systems in some cases. This reduction is based on the use of the Schur product, popular in the field of cryptanalysis (see [5,7,13,14], [17]-[19], [25,27]). For some fixed parameters of the generalized Reed–Solomon codes, Reed–Muller codes, and Goppa codes it was experimentally shown that with high probability, cryptanalysis of Sidel'nikov-type systems can be reduced to cryptanalysis of McEliece-type systems based on Reed–Solomon codes and Reed–Muller codes. Note that for these codes effective cryptanalytic algorithms are known (see [3]-[6]). A theoretical estimate of the probability of fulfilling the equality (14) for randomly selected code D from a given family $\mathcal{E}(C_1, ..., C_u)$ seems to be an actual task. We think that our approach can be applied not only for splitting merged codes of the same dimension but also for splitting codes of different dimensions. In particular, this approach can help to clarify the resistance of a cryptosystem from [28] based on the modification of $(u, u+v)$-construction.

We believe that the cryptosystem $\mathrm{SidMod}(C_1, ..., C_u)$ can be modified by replacing the permutation matrix P in the key (23) with a matrix of a different structure. Such approach is applied in [29] for a McEliece-type cryptosystem. This can complicate the use of attacks based on the Schur product.

The authors sincerely thank G.A. Kabatiansky for discussing the results and useful recommendations for results presentation.

References

1. Sendrier, N., Tillich, J.P.: Code-Based Cryptography: New Security Solutions Against a Quantum Adversary. ERCIM News. https://hal.archives-ouvertes.fr/hal-01410068/document. Accessed 29 Jan 2020
2. McEliece, R.J.: A public-key cryptosystem based on algebraic. Coding Thv. **4244**, 42–44 (1978)
3. Sidel'nikov, V.M., Shestakov, S.O.: On an Encoding System Constructed on the Basis of Generalized Reed-Solomon Codes. Discrete Math. Algorithms Appl. **2**(4), 439–444 (1992). https://doi.org/10.1515/dma.1994.4.3.191
4. Deundyak, V.M., Druzhinina, M.A., Kosolapov, Y.V.: Modification of the Sidel'nikov-Shestakov cryptanalytic algorithm for generalized Reed-Solomon codes and its software implementation. University news. North-caucasian region. Tech. Sci. Ser. **4**, 15–19 (2006). [In Russian]
5. Wieschebrink, C.: Cryptanalysis of the Niederreiter Public Key Scheme Based on GRS Subcodes. In: Sendrier, N. (ed.) PQCrypto 2010. LNCS, vol. 6061, pp. 61–72. Springer, Heidelberg (2010). https://doi.org/10.1007/978-3-642-12929-2_5
6. Minder, L., Shokrollahi, A.: Cryptanalysis of the sidelnikov cryptosystem. In: Naor, M. (ed.) EUROCRYPT 2007. LNCS, vol. 4515, pp. 347–360. Springer, Heidelberg (2007). https://doi.org/10.1007/978-3-540-72540-4_20

7. Borodin, M.A., Chizhov, I.V.: Efficiency of attack on the McEliece cryptosystem constructed on the basis of Reed-Muller codes. Discrete Math. Appl. **24**(5), 273–280 (2014). https://doi.org/10.1515/dma-2014-0024
8. Sendrier, N.: Finding the permutation between equivalent linear codes: the support splitting algorithm. IEEE Trans. Inform. Theory **46**(4), 1193–1203 (2000)
9. Sidelnikov, V.M.: Public-key cryptosystem based on binary Reed-Muller codes. Discr. Math. Appl. **4**(3), 191–208 (1994). https://doi.org/10.1515/dma.1994.4.3.191
10. Davletshina, A.M.: Search for equivalent keys of the McEliece-Sidelnikov cryptosystem based on Reed-Muller binary codes. Appl. Discrete Math. **12**, 98–100 (2019). https://doi.org/10.17223/2226308X/12/31
11. Kabatiansky, G., Tavernier, C.: A new code-based cryptosystem via pseudorepetition of codes. In: Proceedings of Sixteenth International Workshop on Algebraic and Combination Coding Theory, pp. 189–191, Svetlogorsk (Kaliningrad region), Russia (2018)
12. Egorova, E., Kabatiansky, G., Krouk, E., Tavernier, C.: A new code-based public-key cryptosystem resistant to quantum computer attacks. J. Phys. Conf. Series **1163**, 1–5 (2019). https://doi.org/10.1088/1742-6596/1163/1/012061
13. Deundayk, V.M., Kosolapov, Y.V.: The use of the direct sum decomposition algorithm for analyzing the strength of some McEliece-type cryptosystems. Series Math. Mod. Progr. Comp. Soft. **12**(3), 89–101 (2019). https://doi.org/10.14529/mmp190308
14. Deundyak, V.M., Kosolapov, Yu.V.: On the strength of asymmetric code cryptosystems based on the merging of generating matrices of linear codes. In: Proceedings of XVI International Symposium Problems of Redundancy in Information and Control Systems, pp. 143–148, HSE, Moscow, Russia (2019)
15. Randriambololona, H.: On products and powers of linear codes under componentwise multiplication, http://arxiv.org/abs/1312.0022. Accessed 21 Jan 2020
16. Cascudo, I., Cramer, R., Mirandola, D., Zemor, G.: Squares of Random Linear Codes. IEEE Trans. Inform. Theory **61**(3), 1159–1173 (2015). https://doi.org/10.1109/TIT.2015.2393251
17. Vysotskaya, V.: Reed-Muller code square and equivalence classes of secret keys of the McEliece-Sidelnikov cryptosystem. Appl. Discrete Math. **10**, 66–68 (2017). https://doi.org/10.17223/2226308X/10/28
18. Chizhov, I.V., Borodin, M.A.: Cryptanalysis of the McEliece cryptosystem built on $(k-1)$-subcodes of the Reed-Muller code. Appl. Discrete Math. **9**, 73–75 (2016). https://doi.org/10.17223/2226308X/9/29
19. Gauthier, V., Otmani, A., Tillich, J.-P.: A distinguisher-based attack on a variant of McEliece's cryptosystem based on Reed-Solomon codes,https://arxiv.org/pdf/1204.6459.pdf Accessed 21 Jan 2020
20. Vladut, S., Nogin, D., Tsfasman, M.: Algebraic Geometric Codes: Basic Notions. American Mathematical Society (2007)
21. Slepian, D.: Some further theory of group codes. Bell Syst. Tech. J. **39**, 1219–1252 (1960). https://doi.org/10.1002/j.1538-7305.1960.tb03958.x
22. Mirandola, D., Zemor, G.: Critical pairs for the product singleton bound. IEEE Trans. Inf. Theory **61**(9), 4928–4937 (2015). https://doi.org/10.1109/TIT.2015.2450207
23. Massey, J.L.: Minimal codewords and secret sharing. In: Proceedings of the 6th Joint Swedish-russian International Workshop on Information Theory, Russia, pp. 276–279 (1993)

24. Sendrier, N.: On the concatenated structure of a linear code. Appl. Algebra Eng. Commun. Comp. **9**(3), 221–242 (1998). https://doi.org/10.1007/s002000050104
25. Vysotskaya, V., Chizhov, I.: Equivalence classes of McEliece-Sidelnikov-type cryptosystems. In: Proceedings of Sixteenth International Workshop on Algebraic and Combinatorial Coding Theory, pp. 121–124. Svetlogorsk (Kaliningrad region), Russia (2018)
26. MacWilliams, F.J., Sloane, N.J.A.: The Theory of Error-Correcting Codes. North Holland, Amsterdam (1977)
27. Couvreur, A., Gaborit, P., Gauthier-Umana, V., Otmani, A., Tillich, J.-P.: Distinguisher-based attacks on public-key cryptosystem using Reed-Solomon codes. Des. Codes Cryp. **73**(2), 641–666 (2014). https://doi.org/10.1007/s10623-014-9967-z
28. Krasavin, A.A.: Using the modified $(u|u+v)$ - construction in McEliece cryptosystem. Proc. MIPT **10**(2), 189–191 (2018). [In Russian]
29. Baldi, M., Bianchi, M., Chiaraluce, F., Rosenthal, J., Schipani, D.: Enhanced public key security for the McEliece cryptosystem. J. Cryptol. **29**(1), 1–27 (2014). https://doi.org/10.1007/s00145-014-9187-8

A New Code-Based Cryptosystem

Fedor Ivanov[1(✉)], Grigory Kabatiansky[2], Eugeny Krouk[3], and Nikita Rumenko[1]

[1] National Research University Higher School of Economics, Moscow, Russia
fivanov@hse.ru
[2] Skolkovo - Institute of Science and Technology (Skoltech), Moscow, Russia
[3] National Research University Higher School of Economics, Moscow, Russia

Abstract. Unlike most papers devoted to improvements of code-based cryptosystem, where original Goppa codes are substituted by some other codes, we suggest a new method of strengthening which is code-independent. We show (up to some limit) that the security of the new code-based cryptosystem is much closer to the hardness of maximum likelihood decoding than in the original McEliece cryptosystem.

Keywords: McEliece cryptosystem · Code-based cryptography · Key size reduction · Information-set decoding · Maximum likelihood decoding · Bounded distance decoding

1 Introduction

In 1976 W. Diffie and M. Hellman proposed the concept of public-key cryptography concept [1]. To construct this public-key cryptosystem one needs to construct a one-way trap-door function. To achieve this, a hard computational problem should be selected, which nevertheless has simple solutions in some special cases. It is supposed that an eavesdropper who desires to "break" the system, i.e. compute the correspondent plaintext from a given ciphertext, has to solve this hard problem, while a legitimate user, using the corresponding private key, obtains the simple special instance of the hard problem and solves it for decryption.

However, to break the system one may not search for a solution to the hard problem being used, but tries to recover hidden secrets or to construct an equivalent system that produces the same encryption-decryption instead. If the construction of an equivalent system is computationally feasible, this leads to breaking the system without solving the initial hard problem. Such an attack on Merkle-Hellman cryptosystem [2] was given by A. Shamir [3], and in code-based cryptography the most famous analogous example of attack was given in [4] to break McEliece cryptosystem [5] based on modified Reed-Solomon codes proposed in [6].

McEliece cryptosystem is the oldest and most popular code-based cryptosystem. It was proposed in 1978 and it uses irreducible Goppa codes [7]. It relies

M. Baldi et al. (Eds.): CBCrypto 2020, LNCS 12087, pp. 41–49, 2020.
https://doi.org/10.1007/978-3-030-54074-6_3

on NP-hardness of *maximum likelihood decoding* (*MLD* for short) for general linear codes, i.e., the hardness of finding the nearest codeword regarding the Hamming distance for a given received vector [24]. Since there are some classes of codes such as Reed-Solomon (RS), Bose–Chaudhuri–Hocquenghem (BCH), Goppa, Low-Density Parity-Check (LDPC) codes that have polynomial-time decoding algorithms, they can be used in the construction of the corresponding trap-door function. The main idea underlying in the McEliece cryptosystem is to hide a given structured code with a simple decoding algorithm (secret key), hence presenting it as a random code (open key) for which a simple decoder is unknown. The main point of our improvement is the following. The security of the McEliece cryptosystem is not based on the NP-hardness of the MLD problem, since in the frame of the McEliece cryptosystem only errors of weight up to $d/2$ must be corrected, where d is the minimal code distance of the underlying code. Such algorithms are called *half minimal distance decoding*, or *HMD decoding*. Note that it is unknown if HMD decoding is NP-hard (or not). The best known estimates for the complexity of HMD decoding can be found in [9,10]. We hence propose a new cryptosystem, that relies more on the hardness of the MLD problem than the original McEliece cryptosystem. For the best estimates of the complexity of ML (i.e., minimum distance) decoding see [11].

There is no known effective quantum algorithm to break the McEliece cryptosystem but nevertheless it gains no wide practical usage mainly because of the very large size of its public key. For example, in the original paper by McEliece [5] the public key has size of order 250 Kbits.

There were many attempts to attack or to improve the original McEliece cryptosystem, see [12]. The main idea for improvements is to substitute the original Goppa code that is used in McEliece cryptosystem with some other code with a specific structure that allows to reduce the key size. For instance in [13] Goppa codes were substituted by subfield subcodes of quasi-cyclic generalized Reed-Solomon codes. Similar instances based on QC-LDPC codes and LDGM codes (Low-Density Generator Matrices) were proposed in [15–21].

Also it should be mentioned that there are some frameworks for code-based cryptography, where authors do not only exchange the secret code within the McEliece cryptosystem. For instance, see [22–25].

The new code-based cryptosystem proposed in this paper forces the eavesdropper to correct seemingly random errors and gives another way to shrink public key sizes due to shorter codes.

The structure of the paper is the following: we start from the standard McEliece cryptosystem, then we describe a "prototype" cryptosystem which has nice mathematical structure but unfortunately provides gain compared with the original McEliece, and finally we propose a new scheme with better parameters.

2 McEliece Cryptosystem

2.1 Design

In the following we recall how the McEliece cryptosystem works. There are two users Alice and Bob, where Bob wants to send a k bit message m to Alice. Alice takes a $k \times n$ generator matrix G of some linear (n, k)-code C with the minimal code distance $d(C) \geq 2t + 1$, which has an efficient decoding algorithm Φ, correcting t errors. The matrix G is a secret, known only to Alice, and the code C is called the *secret code*.

Alice constructs a *public* matrix $G_{pub} = SGP$, where a $k \times k$ nonsingular matrix S and a $n \times n$ permutation matrix P are chosen randomly from the corresponding ensembles and they are also keeping as secrets.

Bob sends to Alice the following *ciphertext* y

$$y = mG_{pub} + e, \tag{1}$$

where e is a vector of weight t which is *randomly* generated by Bob. Alice reveals the message m by the following chain of simple calculations:

$$y' := yP^{-1} = mG_{pub}P^{-1} + eP^{-1} = m'G + e', \tag{2}$$

where $m' = mS$, $e' = eP^{-1}$ and $wt(e') = wt(e) = t$, since P is a permutation matrix. Then Alice applies the decoding algorithm Ψ to the vector $y' = m'G+e'$ and receives $\Psi(y') = m'$ and finally finds $m := m'S^{-1}$.

Any other user will deal either with the problem of correcting t errors of a random looking linear code C_{pub} with generator matrix G_{pub} or with the problem of reconstructing the code structure from its public-key matrix, these attacks are called *structural attacks*. In the original paper [5] irreducible Goppa codes [7] were chosen as the family of codes for the scheme. In particular, it was suggested to use Goppa code of length $n = 1024$ dimension $k = 524$ and minimum distance $d = 101$, hence $t = 50$.

Later H. Niederreiter proposed a cryptosystem [6], which is based on solving a syndrom equation and in some sense is dual to the McEliece scheme. These two schemes have equivalent security [26] and we restrict our consideration to the McEliece type schemes.

2.2 Decoding Attacks on McEliece Cryptosystem

An attacker (or eavesdropper) $\mathcal{E}$ve tries to find a vector $\hat{e}$ such that

$$y - \hat{e} \in C_{pub}. \tag{3}$$

If $\hat{e} = e$ then (3) holds and $\mathcal{E}$ finds the message m from $mG_{pub} = y - \hat{e}$.

Note that for $\hat{e} \neq e$ the Eq. (3) does not hold. Indeed, let $y - \hat{e} \in C_{pub}$. Since $y - e \in C_{pub}$ we have that $e - \hat{e} \in C_{pub}$. The *public* code C_{pub} is equivalent to the code C, therefore its distance $d(C_{pub}) \geq 2t + 1$, but $wt(e - \hat{e}) \leq wt(e) + wt(\hat{e}) = 2t$ and hence $\hat{e} = e$.

In the worst case $\mathcal{E}$ve must try $\binom{n}{t}(q-1)^t$ vectors $\hat{e}$ over $\mathbb{F}_q$, and on average it takes half of this value, which is nevertheless a huge number for any reasonable code parameters.

Much more effective is the attack based on *Information Set Decoding* (ISD). This attack was already mentioned in the initial security analysis of McEliece [5] and further developed in numerous papers, see [12] and references there. There are different interpretations and modifications of the initial ISD algorithm. Several different improvements have been proposed, such as ball-collision decoding [12] and improvements based on generalized birthday approaches. For instance, in paper [9] the complexity of ISD was reduced to $\tilde{\mathcal{O}}(2^{0.054n})$ and in [10] the complexity exponent is $\tilde{\mathcal{O}}(2^{0.0494n})$ which is the currently the best result.

In the following we recall the basic properties of ISD algorithms. Goal of ISD algorithms is to recover the message m from a given vector $y = m\hat{G} + e$, where $\hat{G}$ is a generator matrix of an (n, k) code $\hat{C}$ with minimal distance $d \geq 2t + 1$ and $wt(e) \leq t$.

Let I be a k-subset of the coordinates set $[n] := \{1, 2, \dots, n\}$ such that I is an information set of $\hat{C}$ and $\hat{G}_I$ be the submatrix of $\hat{G}$ consisting of columns indexed by I. In the same way let e_I be the vector consisting of coordinates of the vector e indexed by I. ISD algorithms work in the following way:

1. Randomly choose an information set I.
2. Find a codeword $\hat{c}$ such that $\hat{c}_I = y_I$
3. Check if $wt(\hat{c} - y) = t$. If Yes then output the message corresponding to the codeword $\hat{c}$. Else return to Step 1.

Observe that, if one assumes that the support of the error vector is disjoint from the information set, then the corresponding probability P_k that chosen k coordinates are error-free is

$$P_k = \frac{\binom{n-t}{k}}{\binom{n}{k}} = \frac{\binom{n-k}{t}}{\binom{n}{t}}, \tag{4}$$

and hence the the average number of required iterations is of order $\frac{\binom{n}{t}}{\binom{n-k}{t}}$, which is significantly less than the complexity of the brute-force attack.

In the next section we will describe a "prototype" of a new cryptosystem.

3 The "Prototype" Code-Based Cryptosystem

Let C be a linear (n, k)-code with the minimum distance $d(C) \geq 2t + 1$, which has an efficient decoding algorithm Φ, correcting t errors. We also assume that C belongs to some rather big family of codes (like Goppa codes in the original McEliece cryptosystem). Alice takes $k \times n$ generator matrix G of this code C. The matrix G as well as the code C are secrets, known only to Alice, and we call code C is called the *secret code*.

Alice constructs *two* public matrices, namely $G_{pub} = GM$, where M is a randomly chosen $n \times n$ non-singular matrix, and $E_{pub} = (C_n + P)M$, where P is a randomly chosen $n \times n$ permutation matrix and C_n is $n \times n$ matrix which rows are codewords of the code C, i.e. $C_n = UG$ for some random $n \times k$ matrix U. We put some additional restriction on joint choice of matrices P and C_n later, in order to avoid structural attacks. Matrices P and C_n are kept secret.

Bob sends to Alice the following *ciphertext* y

$$y = mG_{pub} + eE_{pub} = (mG + e(C_n + P))M, \tag{5}$$

where e is a vector of weight t *randomly* generated by Bob. Alice reveals the message m by the following chain of calculations:

$$y' := yM^{-1} = mG + e(C_n + P) = m'G + e', \tag{6}$$

where $m' = m + eU,\ \ e' = eP$ and $wt(e') = wt(e) = t$, since P is a permutation matrix, then Alice applies the decoding algorithm Ψ to the vector $y' = m'G + e'$, which outputs the *error vector* e', hence Alice knows $e = e'P^{-1}$ and finally finds m, for instance, from $mG_{pub} = y - eE_{pub}$, see (5).

3.1 First Attack or Why Matrix E_{pub} Must be Singular

Let us show that if the matrix E_{pub} is non-singular then the new scheme can be attacked the same way as the original McEliece scheme. Indeed, if $E_{pub} = (C_n + P)M$ is non-singular then Eve can compute vector $\tilde{y} := yE_{pub}^{-1}$. Hence, according to (5),

$$\tilde{y} = (mG_{pub} + eE_{pub})E_{pub}^{-1} = mG(C_n + P)^{-1} + e = m\tilde{G} + e, \tag{7}$$

where $\tilde{G} = G(C_n + P)^{-1}$ can be considered as a generator matrix of some linear (n, k)-code $\tilde{C}$. It is easy to see that the Eq, (7) cannot have more than one solution for a given $\tilde{y}$. Hence, the code $\tilde{C}$ has distance at least $2t + 1$ and ISD algorithms can be applied. Moreover, we shall show that codes C and $\tilde{C}$ are permutation equivalent and thus to break our scheme in the case where E_{pub} is invertible is the same as to break the McEliece cryptosystem.

Remark 1. To prove that codes C and $\tilde{C}$ are equivalent recall that the rows of the matrix C_n are of the form uG (for some k-tuple u) since they are vectors of the code C and it would be convenient to represent matrix C_n as UG, where U is the corresponding $n \times k$ matrix.

Let us start from the following obvious equality

$$G(I_n + P^{-1}UG) = (I_k + GP^{-1}U)G,$$

and hence

$$(I_k + GP^{-1}U)^{-1}G = G(I_n + P^{-1}UG)^{-1}. \tag{8}$$

By the definition $\tilde{G} = G(UG + P)^{-1}$ and thus

$$\tilde{G} = G(P(P^{-1}UG + I_n))^{-1} = G(P^{-1}UG + I_n)^{-1}P^{-1} = (I_k + GP^{-1}U)^{-1}GP^{-1}.$$

Hence, we proved that $\tilde{G} = (I_k + GP^{-1}U)^{-1}GP^{-1}$, and therefore codes C and $\tilde{C}$ are permutation equivalent (if matrix $(I_k + GP^{-1}U)^{-1}$ exists).

3.2 How to Make Matrix E_{pub} Singular

The matrix $E_{pub} = (C_n + P)M$ is singular iff matrix $C_n + P$ is singular since matrix M is non-singular. Let us show how to construct many singular matrices of the form $C_n + P$. Note that w.l.o.g. we can restrict our consideration to the case $C_n + I_n$ and then transform the obtained singular matrices to the desired ones of the form $\breve{C}_n + P$, where $\breve{C}_n = C_n P$.

Let us first, for simplicity, consider the binary case. Let $\mathbf{c} = (c_1, \ldots, c_n) \in C$ be a codeword of the Hamming weight w and let $c_{j_1}, \ldots, c_{j_w}$ be its w nonzero coordinates. Construct rows $\mathbf{c}^i$ of C_n in the following way: rows not indexed by $J = supp(c)$ will be taken randomly, and the rest of the rows are chosen in such a way that

$$\sum_{j \in J} \mathbf{c}^j = \mathbf{c}. \tag{9}$$

Denote by $B_n = C_n + I_n$ and let $\mathbf{b}^i = \mathbf{c}^i + \delta_i$ be the i-th row of the matrix B, where $\delta_i = (\delta_{i,1}, \ldots, \delta_{i,n})$ and $\delta_{i,j}$ is the Kronecker delta. Then $\sum_{j \in J} \mathbf{b}^j = \mathbf{0}$ and thus the matrix $B_n = C_n + I_n$ is singular.

In the general case one should replace Eq. (9) on

$$\sum_{j \in J} c_j \mathbf{c}^j = -\mathbf{c} \tag{10}$$

and then $\sum_{j \in J} c_j \mathbf{b}^j = \mathbf{0}$ and thus the matrix $B_n = C_n + I_n$ is singular.

Obviously the number of solutions of Eq. (9) equals to $q^{k(w-1)}$, since say first $w - 1$ vectors $\mathbf{c}^j$ can be chosen as arbitrary codevectors, and the last one is chosen uniquely according to (9). Hence the total number of matrices constructed according to (9) for a given nonzero codeword c equals to $q^{k(n-1)}$, among total number q^{kn} $n \times n$ matrices, whose rows are vectors of the code C.

3.3 Second Attack Based on Parity-Check Matrix H_{pub}

Unfortunately, there is another attack which shows that the "prototype" cryptosystem can be successfully attacked at least by ISD algorithms.

Namely, Eve computes a parity-check matrix H_{pub} for the generator matrix G_{pub}, i.e. $G_{pub}H_{pub}^T = H_{pub}G_{pub}^T = 0$. Let H be a parity-check matrix for the code C, i.e. $GH^T = 0$. Then it is easy to see that $H_{pub}^T = M^{-1}H^T S^T$, where S

is some non-singular $r \times r$-matrix and $r = n - k$. After that Eve multiplies both parts of Eq. (5), where $C_n = UG$, from the right side with H_{pub}^T and receives

$$\tilde{y} := yH_{pub}^T = (mG + e(UG + P))MM^{-1}H^T S^T = ePH^T S^T = e\tilde{H}^T. \quad (11)$$

Hence (11) is a usual syndrome equation for the code $\tilde{C}$ with parity-check matrix $\tilde{H} = SHP^T$. Since obviously codes C and $\tilde{C}$ are permutation equivalent and the "prototype" cryptosystem is not more secure that the ordinary McEliece system but even worse its public keys are at least twice as large.

4 The New Code-Based Cryptosystem

In order to improve the "prototype" system we shall make the structure of the matrix E_{pub} more complicated. Namely, let $E_{pub} = WD(C_n + P)M$, where $(C_n + P)M$ is the same as for the prototype, D is a randomly chosen $n \times n$ diagonal matrix with t non-zero elements on the diagonal, and W is random $n \times n$ non-singular matrix. Alice forms two *public* matrices: $k \times n$ matrix $G_{pub} = GM$ and $n \times n$ matrix $E_{pub} = WD(C_n + P)M$.

Bob sends to Alice the following *ciphertext* y

$$y = mG_{pub} + eE_{pub} = (mG + eWD(C_n + P))M, \quad (12)$$

where e is a vector *randomly* generated by Bob. Let us stress that the vector e does not bear any restriction on its weight. Recall that C_n can be represented as $C_n = UG$, where U is the appropriate $n \times k$ matrix, and Alice reveals the message m by the following chain of calculations:

$$y' := yM^{-1} = mG + eWD(UG + P) = m'G + e'P, \quad (13)$$

where $m' = m + eWDU$, $e' = (eW)D$. Note that $wt(e') \leq t$, since D is a diagonal matrix of "weight" t. Then as for the prototype Alice applies the decoding algorithm Ψ to the vector $y' = m'G + e''$, where $e'' = e'P$, which outputs "error vector" e''. Hence Alice knows $e' = e''P^{-1}$, thereafter subsequently finds eWD, then $eWDC_n$ and finally finds m, for instance, from $mG_{pub} = y - eE_{pub}$, see (12).

It is straightforward to check that both attacks described for the "prototype" system do not work for the new system. Indeed, matrix E_{pub} has rank t, since matrix D has rank t, and thus the first attack cannot be applied.

For the second attack Eve multiplies both parts of Eq. (12) from the right side on $H_{pub}^T = M^{-1}\tilde{H}^T$, where $\tilde{H}$ is some parity-check matrix of the code C. She receives the following equation

$$yH_{pub}^T = (mG + eWD(UG + P))MM^{-1}\tilde{H}^T = eWDP\tilde{H}^T = (eWDP)\tilde{H}^T \quad (14)$$

which is a usual syndrome equation for a code with parity-check matrix $\tilde{H}$ and hence Eve can find vector $eWDP$ of weight t but she cannot "extract" from it the vector e since all three multipliers W, D and P are unknown to her.

The new cryptosystem forces Eve to apply brute-force attacks which have complexity at least $\binom{n}{t}$ *trials.*

Consider the following example

Example 1. Let C be an irreducible Goppa code of length $n = 256$ and rate $R = 1/2$, i.e. with $k = 128$ and $t = 16$. Then the number of trials is $\binom{16}{256} > 2^{100}$ and the public key length is $128 \times 256 + 256^2 = 3 \times 2^{15}$.

5 Conclusion

In this paper we considered a new modification of the well-known McEliece cryptosystem in which we transform an error vector of weight t (or $\leq t$) to an error vector of arbitrary weight.

References

1. Diffie, W., Hellman, M.: New directions in cryptography. IEEE Trans. Inf. Theory **22**(6), 644–654 (1976)
2. Merkle, R., Hellman, M.: Hiding information and signatures in trapdoor knapsacks. IEEE Trans. Inf. Theory **24**(5), 525–530 (1978)
3. Shamir, A.: A polynomial-time algorithm for breaking the basic Merkle-Hellman cryptosystem. IEEE Trans. Inf. Theory **30**(5), 699–704 (1984)
4. Sidelnikov, V.M., Shestakov, S.O.: On encryption based on generalized reed solomon codes. Discrete Math. Appl. **2**(4), 439–444 (1992)
5. McEliece, R.J.: A public-key cryptosystem based on algebraic Coding Theory. DSN Progress Report 42–44, pp. 114–116 (1978)
6. Niederreiter, H.: Knapsack-type cryptosystems and algebraic coding theory. Prob. Control Inf. Theory **15**, 159–166 (1986)
7. Goppa, V.D.: A new class of linear correcting codes. Problemy Peredachi Informatsii **6**(3), 24–30 (1970)
8. Berlekamp, E.R., McEliece, R.J., van Tilborg, H.C.A.: On the inherent intractability of certain coding problems. IEEE Trans. Inform. Theory **24**, 384–386 (1978)
9. May, A., Meurer, A., Thomae, E.: Decoding random linear codes in $\tilde{\mathcal{O}}(2^{0.054n})$. In: Lee, D.H., Wang, X. (eds.) ASIACRYPT 2011. LNCS, vol. 7073, pp. 107–124. Springer, Heidelberg (2011). https://doi.org/10.1007/978-3-642-25385-0_6
10. Becker, A., Joux, A., May, A., Meurer, A.: Decoding random binary linear codes in $2^{n/20}$: how $1 + 1 = 0$ improves information set decoding. In: Pointcheval, D., Johansson, T. (eds.) EUROCRYPT 2012. LNCS, vol. 7237, pp. 520–536. Springer, Heidelberg (2012). https://doi.org/10.1007/978-3-642-29011-4_31
11. Barg, A., Krouk, E., van Tilborg, H.: On the complexity of minimum distance decoding of long linear codes. IEEE Trans. Inf. Theory **45**(5), 1392–1405 (1999)
12. Bernstein, D.J., Lange, T., Peters, C.: Attacking and defending the McEliece cryptosystem. In: Buchmann, J., Ding, J. (eds.) PQCrypto 2008. LNCS, vol. 5299, pp. 31–46. Springer, Heidelberg (2008). https://doi.org/10.1007/978-3-540-88403-3_3

13. Berger, T.P., Cayrel, P.-L., Gaborit, P., Otmani, A.: Reducing key length of the McEliece cryptosystem. In: Preneel, B. (ed.) AFRICACRYPT 2009. LNCS, vol. 5580, pp. 77–97. Springer, Heidelberg (2009). https://doi.org/10.1007/978-3-642-02384-2_6
14. Von Maurich, I., Güneysu, T.: Lightweight code-based cryptography: QC-MDPC McEliece encryption on reconfigurable devices, In 2014 Design, Automation and Test in Europe Conference and Exhibition (DATE), pp. 1–6 (2014)
15. Baldi, M., Chiaraluce, F., Garello, R., Mininni, F.: Quasi-cyclic low-density parity-check codes in the McEliece cryptosystem. In: 2007 IEEE International Conference on Communications, pp. 951–956 (2007)
16. Baldi, M.: LDPC codes in the McEliece cryptosystem: attacks and countermeasures, In: NATO Science for Peace and Security Series–D: Information and Communication Security. LNCS, vol. 23, pp. 160–174 (2009)
17. Baldi, M., Bodrato, M., Chiaraluce, F.: A new analysis of the McEliece cryptosystem based on QC-LDPC codes. In: Ostrovsky, R., De Prisco, R., Visconti, I. (eds.) SCN 2008. LNCS, vol. 5229, pp. 246–262. Springer, Heidelberg (2008). https://doi.org/10.1007/978-3-540-85855-3_17
18. Baldi, M., Bambozzi, F., Chiaraluce, F.: On a family of circulant matrices for quasi-cyclic low-density generator matrix codes. IEEE Trans. Inf. Theory **57**(9), 6052–6067 (2011)
19. Baldi, M., Bianchi, M., Chiaraluce, F.: Security and complexity of the McEliece cryptosystem based on quasi-cyclic low-density parity-check codes. IET Inf. Secur. **7**(3), 212–220 (2013)
20. Baldi, M., Bianchi, M., Chiaraluce, F.: Optimization of the parity-check matrix density in QC-LDPC code-based McEliece cryptosystems. In: Workshop on Information Security Over Noisy and Lossy Communication Systems (IEEE ICC 2013) (2013)
21. Misoczki, R., Tillich, J.P., Sendrier, N., Barreto, P.S.: MDPC-McEliece: new McEliece variants from moderate density parity-check codes. In: 2013 IEEE International Symposium on Information Theory, pp. 2069–2073 (2013)
22. Alekhnovich, M.: More on average case vs approximation complexity. In: 44th Annual IEEE Symposium on Foundations of Computer Science, Proceedings, pp. 298–307 (2003)
23. Baldi, M., Bianchi, M., Chiaraluce, F., Rosenthal, J., Schipani, D.: A variant of the McEliece cryptosystem with increased public key security. In: Proceedings of WCC 2011 - Seventh Workshop on Coding and Cryptography, no. 7, pp. 173–182. HAL-Inria (2011)
24. Berlekamp, E., McEliece, R.J., Van Tilborg, H.: On the inherent intractability of certain coding problems. IEEE Trans. Inf. Theory **24**(3), 384–386 (1978)
25. Khathuria, K., Rosenthal, J., Weger, V.: Encryption scheme based on expanded Reed-Solomon codes. Advances in Mathematics of Communications (2019)
26. Li, Y.X., Deng, R.H., Wang, X.M.: On the equivalence of McEliece's and Niederreiter's public-key cryptosystems. IEEE Trans. Inf. Theory **40**(1), 271–273 (1994)

On Constant-Time QC-MDPC Decoders with Negligible Failure Rate

Nir Drucker[1,2(✉)], Shay Gueron[1,2], and Dusan Kostic[3]

[1] University of Haifa, Haifa, Israel
drucker.nir@gmail.com
[2] Amazon, Seattle, USA
[3] EPFL, Lausanne, Switzerland

Abstract. The QC-MDPC code-based KEM Bit Flipping Key Encapsulation (BIKE) is one of the Round-2 candidates of the NIST PQC standardization project. It has a variant that is proved to be IND-CCA secure. The proof models the KEM with some black-box ("ideal") primitives. Specifically, the decapsulation invokes an ideal primitive called "decoder", required to deliver its output with a negligible Decoding Failure Rate (DFR). The concrete instantiation of BIKE substitutes this ideal primitive with a new decoding algorithm called "Backflip", that is shown to have the required negligible DFR. However, it runs in a variable number of steps and this number depends on the input and on the key. This paper proposes a decoder that has a negligible DFR and also runs in a fixed (and small) number of steps. We propose that the instantiation of BIKE uses this decoder with our recommended parameters. We study the decoder's DFR as a function of the scheme's parameters to obtain a favorable balance between the communication bandwidth and the number of steps that the decoder runs. In addition, we build a constant-time software implementation of the proposed instantiation, and show that its performance characteristics are quite close to the IND-CPA variant. Finally, we discuss a subtle gap that needs to be resolved for every IND-CCA secure KEM (BIKE included) where the decapsulation has nonzero failure probability: the difference between average DFR and "worst-case" failure probability per key and ciphertext.

Keywords: BIKE · QC-MDPC codes · IND-CCA · Constant-time algorithm · Constant-time implementation

1 Introduction

BIKE [3] is a code-based Key Encapsulation Mechanism (KEM) using Quasi-Cyclic Moderate-Density Parity-Check (QC-MDPC) codes. It is one of the Round-2 candidates of the NIST PQC Standardization Project [16]. BIKE submission includes three variants (BIKE-1, BIKE-2, and BIKE-3) with three security levels for each one. Hereafter, we focus mainly on BIKE-1, at its Category 1 (as defined by NIST) security level.

M. Baldi et al. (Eds.): CBCrypto 2020, LNCS 12087, pp. 50–79, 2020.
https://doi.org/10.1007/978-3-030-54074-6_4

The decapsulation algorithm of BIKE invokes an algorithm that is called a decoder. The decoder is an algorithm that, given prescribed inputs, outputs an error vector that can be used to extract a message. There are various decoding algorithms and different choices yield different efficiency and DFR properties.

QC-MDPC Decoding Algorithms. We briefly describe the evolution of several QC-MDPC decoding algorithms. All of them are derived from the Bit-Flipping algorithm that is commonly attributed [10]. The Round-1 submission of BIKE describes the "One-Round" decoder. This decoder is indeed implemented in the accompanying reference code [3]. The designers of the constant-time Additional implementation [7] of BIKE Round-1 chose to use a different decoder named "Black-Gray"[1], with rationale as explained in [6]. The study presented in [20] explores two additional variants of the Bit-Flipping decoder: a) a parallel algorithm similar to that of [10], which first calculates some thresholds for flipping bits, and then flips the bits in all of the relevant positions, in parallel. We call this decoder the "Simple-Parallel" decoder; b) a "Step-by-Step" decoder (an enhancement of the "in-place" decoder described in [8]). It recalculates the threshold every time that a bit is flipped.

The Round-2 submission of BIKE uses the One-Round decoder (of Round-1) and a new variant of the Simple-Parallel decoder. The latter introduces a new trial-and-error technique called Time-To-Live (TTL). It is positioned as a derivative of the decoders in [20]. All of these decoders have some nonzero probability to fail in decoding a valid input. The average of the failure probability over all the possible inputs (keys and messages) is called Decoding Failure Rate (DFR). The KEMs of Round-1 BIKE were designed to offer IND-CPA security and to be used only with ephemeral keys. They had an approximate DFR of 10^{-7}, which is apparently tolerable in real systems. As a result, they enjoyed acceptable bandwidth and performance. Round-2 submission presented new variants of BIKE KEMs that provide IND-CCA security. Such KEMs can be used with static keys. The IND-CCA BIKE is based on three changes over the IND-CPA version: a) a transformation (called $FO^{\not\perp}$) applied to the key generation, encapsulation and decapsulation of the original IND-CPA flows (see [3][Section 6.2.1]); b) adjusted parameters sizes; c) invoking the Backflip decoder in the decapsulation algorithm.

Our Contribution

- We define Backflip$^+$ decoder as the variant of Backflip that operates with a fixed X_{BF} number of iterations (for some X_{BF}). We also define the Black-Gray decoder that runs with a given number of iterations X_{BG} (for some X_{BG}). Subsequently, we analyze the DFR of these decoders as a function of X_{BF} and X_{BG} and the block size (which determines the communication bandwidth of the KEM). The analysis finds a new set of parameters where Backflip$^+$ with $X_{BF} = 8, 9, 10, 11, 12$ and the Black-Gray decoder with

[1] This decoder appears in the pre-Round-1 submission "CAKE" (the BIKE-1 ancestor). It is due to N. Sendrier and R. Misoczki. The decoder was adapted to use the improved thresholds published in [3].

$X_{BG} = 3, 4, 5$ have an estimated average DFR of 2^{-128}. This offers multiple IND-CCA proper BIKE instantiation options.

- We build an optimized constant-time implementation of the new BIKE CCA flows together with a constant-time implementation of the decoders. This facilitates a performance comparison between the Backflip$^+$ and the Black-Gray decoders. All of our performance numbers are based *only* on constant-time implementations. The comparison leads to interesting results. The Backflip$^+$ decoder has a better DFR than the Black-Gray decoder if both of them are allowed to have a very large (practically unlimited) X_{BF} and X_{BG} values. These values do not lead to practical performance. However, for small X_{BF} and X_{BG} values that make the performance practical and DFR acceptable, the Black-Gray decoder is faster (and therefore preferable).
- The BIKE CCA flows require higher bandwidth and more computations compared to the original CPA flows, but the differences as measured on x86-64 architectures are not very significant. Table 1 summarizes the trade-off between the BIKE-1 block size (r), the estimated DFR and the performance of BIKE-1 decapsulation (with IND-CCA flows) using the Black-Gray decoder. It provides several instantiations/implementations choices. For example, with $X_{BG} = 4$ iterations and targeting a DFR of 2^{-64} (with $r = 11,069$ bits) the decapsulation with the Black-Gray decoder consumes 4.81M cycles. With a slightly higher $r = 11,261$ the decoder can be set to have only $X_{BG} = 3$ iterations and the decapsulation consumes 3.76M cycles.

Table 1. The BIKE-1 Level-1 block size r (in bits) for which the Black-Gray decoder achieves a target DFR with a specified number of iterations, and the decapsulation performance (in cycles; the precise details of the platform are provided in Sect. 5). A DFR of 2^{-128} is required for the IND-CCA KEM. The IND-CPA used with ephemeral keys can settle with higher DFR.

DFR		3 iterations	4 iterations	5 iterations
$2^{-23} \approx 10^{-7}$	r	10,259	10,163	10,141
	Cycles	3.50M	4.52M	5.53M
$2^{-30} \approx 10^{-9}$	r	10,427	10,331	10,301
	Cycles	3.52M	4.56M	5.63M
$2^{-40} \approx 10^{-12}$	r	10,667	10,589	10,501
	Cycles	3.55M	4.63M	5.69M
2^{-64}	r	11,261	11,069	11,003
	Cycles	3.76M	4.81M	5.96M
2^{-128}	r	12,781	12,437	12,373
	Cycles	4.06M	5.22M	6.47M

- The $FO^{\not\perp}$ transformation from QC-MDPC McEliece Public Key Encryption (PKE) to BIKE-1 IND-CCA relies on the assumption that the underlying

PKE is δ-correct [12] with $\delta = 2^{-128}$. The relation between this assumption and the (average) DFR that used in [3] is not yet addressed. We identify this gap and illustrate some of the remaining challenges.

The paper is organized as follows. Section 2 offers background, notation and surveys some QC-MDPC decoders. In Sect. 3 we define and clarify subtle differences between schemes using idealized primitives and concrete instantiations of the schemes. In Sect. 4 we explain the method used for estimating the DFR. We explain the challenges and the techniques that we used for building a constant-time implementation of IND-CCA BIKE in Appendix B. Section 5 reports our results for the DFR and block size study, and also the performance measurements of the constant-time implementations. The gap between the estimated DFR and the δ-correctness needed for IND-CCA BIKE is discussed in Sect. 6. Section 7 concludes this paper with several concrete proposals and open questions.

2 Preliminaries and Notation

Let $\mathbb{F}_2$ be the finite field of characteristic 2. Let $\mathcal{R}$ be the polynomial ring $\mathbb{F}_2[X]/\langle X^r - 1\rangle$. For every element $v \in \mathcal{R}$ its Hamming weight is denoted by $wt(v)$. The length of a vector w is denoted by $|w|$. Polynomials in $\mathcal{R}$ are viewed interchangeably also as square circulant matrices in $\mathbb{F}_2^{r\times r}$. For a matrix $H \in \mathbb{F}_2^{r\times r}$ let h_j denote its j-th column written as a row vector. We denote null values and protocol failures by $\perp$. Uniform random sampling from a set W is denoted by $w \xleftarrow{\$} W$. For an algorithm A, we denote its output by $out = A()$ if A is deterministic, and by $out \leftarrow A()$ otherwise. Hereafter, we use the notation $x.ye{-}z$ to denote the number $(x + \frac{y}{10}) \cdot 10^{-z}$.

2.1 BIKE-1

The computations of BIKE-1-(CPA/CCA) are executed over $\mathcal{R}$, where r is a given parameter. Let w and t be the weights of (h_0, h_1) in the secret key $h = (h_0, h_1, \sigma_0, \sigma_1)$ and the errors vector $e = (e_0, e_1)$, respectively. Denote the public key, ciphertext, and shared secret by $f = (f_0, f_1)$, $c = (c_0, c_1)$, and k, respectively. As in [3], we use **H**, **K** to denote hash functions. Currently, the parameters of BIKE-1-CPA for NIST Level-1 are $r = 10{,}163$, $|f| = |c| = 20{,}326$ and for BIKE-1-CCA are $r = 11{,}779$, $|f| = |c| = 23{,}558$. In both cases, $|k| = 256$, $w = 142$, $d = w/2 = 71$ and $t = 134$. Figure 1 shows the BIKE-1-CPA and BIKE-1-CCA flows [3], see details therein.

2.2 The IND-CCA Transformation

Round-2 BIKE submission [3] uses the $FO^{\not\perp}$ conversion ([12] which relies on [9]) to convert the QC-MDPC McEliece PKE into an IND-CCA KEM BIKE-1-CCA. The submission claims that the proof results from [12][Theorems 3.1 and 3.4[2]].

[2] Theorems 3.1 and 3.4 appear only in the ePrint version [13] of [12]. In [12] they appear as Theorems 1 and 4, respectively.

	BIKE-1 IND-CPA	BIKE-1 IND-CCA
Key generation	$h_0, h_1 \xleftarrow{\$} \mathcal{R}$ of odd weight $wt(h_0) = wt(h_1) = w/2$	
	-	$\sigma_0, \sigma_1 \xleftarrow{\$} \mathcal{R}$
	$g \xleftarrow{\$} \mathcal{R}$ of odd weight (so $wt(g) \approx r/2$) $(f_0, f_1) = (gh_1, gh_0)$	
Encapsulation	$m \xleftarrow{\$} \mathcal{R}$	
	$e_0, e_1 \xleftarrow{\$} \mathcal{R}$	$(e_0, e_1) = \mathbf{H}(mf_0, mf_1)$
	where $wt(e_0) + wt(e_1) = t$	
	$(c_0, c_1) = (mf_0 + e_0, mf_1 + e_1)$	
	$k = \mathbf{K}(e_0, e_1)$	$k = \mathbf{K}(mf_0, mf_1, c_0, c_1)$
Decapsulation	Compute the syndrome $s = c_0h_0 + c_1h_1$ $(e_0', e_1') \leftarrow \text{decode}(s, h_0, h_1)$ If $wt\left((e_0', e_1')\right) \neq t$ or decoding failed then	
	return $\perp$	$k = \mathbf{K}(\sigma_0, \sigma_1, c)$
	else	
	$k = \mathbf{K}(e_0', e_1')$	$k = \mathbf{K}(c_0 + e_0', c_1 + e_1', c_0, c_1)$

Fig. 1. BIKE-1 IND-CPA/IND-CCA flows (full details are given in [3]).

These theorems use the term δ-correct PKEs. For a finite message space M, a PKE is called δ-correct when[3]

$$\mathbb{E}\left[\max_{m \in M} Pr\left[Decrypt(sk, c) \neq m \mid c \leftarrow Encrypt(pk, m)\right]\right] \leq \delta \quad (1)$$

a KEMs is δ-correct if

$$Pr\left[Decaps(sk, c) \neq K | (sk, pk) \leftarrow Gen(), (c, K) \leftarrow Encaps(pk)\right] \leq \delta \quad (2)$$

2.3 QC-MDPC Decoders

The QC-MDPC decoders discussed in this paper are variants of the Bit Flipping decoder [10] presented in Algorithm 1. They receive a parity check matrix $H \in \mathbb{F}_2^{r \times n}$ and a vector $c = mf + e$ as input[4]. Here, $c, mf, e \in \mathbb{F}_2^n$, mf is a codeword (thus, $H(mf)^T = 0$) and e is an error vector with small weight. The algorithm calculates the syndrome $s = eH^T$ and subsequently extracts e' from s. The goal of the Bit Flipping algorithm is to have e' such that $e' = e$.

Algorithm 1 consists of four steps: I) calculate some static/dynamic threshold (th) based on the syndrome (s) and the error (e) weights; II) compute the number of unsatisfied parity check equations (upc_i) for a given column $i \in \{0, \dots, n-1\}$; III) Flip the error bits in the positions where there are more unsatisfied parity-check equations than the calculated threshold; IV) Recompute the syndrome.

[3] In BIKE-1, the secret key (sk) and public key (pk) are h and f, respectively.

[4] In BIKE-1, $n = 2r$, the parity-check matrix H is formed by the two circulant blocks (h_0, h_1), the vectors c, e, and f are defined as $c = (c_0, c_1)$, $e = (e_0, e_1)$, and $mf = (m \cdot f_0, m \cdot f_1)$.

We refer to Algorithm 1 as the Simple-Parallel decoder. The Step-By-Step decoder inserts Steps 4, 9 into the "`for`" loop (Step 5), i. e., it recalculate the threshold and the syndrome for every bit. The One-Round decoder starts with one iteration of the Simple-Parallel decoder, and then switches to the Step-by-Step decoder mode of operation.

Algorithm 1. e=BitFlipping(c, H)

Input: Parity-check matrix $H \in \mathbb{F}_2^{r \times n}$, $c \in \mathbb{F}_2^n$, maxIter (maximal # of iterations), u maximal syndrome weight
Output: The error $e \in \mathbb{F}_2^n$
Exception: "decoding failure" return $\perp$

1: **procedure** BitFlipping(c, H)
2: s $= Hc^T$, e = 0, itr = 0
3: **while** ($wt(s) > u$) and (itr < maxIter) **do**
4: th = computeThreshold(s,e) ▷ Step I
5: **for** i in $0 \dots n-1$ **do**
6: Compute upc_i ▷ Step II
7: **if** $upc_i > th$ **then** $e[i] = e[i] \oplus 1$ ▷ Step III
8: s $= H(c^T + e^T)$ ▷ Step IV
9: itr = itr + 1
10: **if** itr = maxIter **then**
11: **return** $\perp$
12: **else**
13: **return** e

The Black-Gray decoder (in the additional code [7]) and the Backflip decoder [3] use a more complex approach. Similar to the Simple-Parallel decoder, they operate on the error bits in parallel. However, they add a step that re-flips the error bits according to some estimation.

The "`while`" loop of an iteration of the Black-Gray decoder consists of: 1) Perform 1 iteration of the Simple-Parallel decoder and define some bits position candidates that should be reconsidered (i. e., bits that were mistakenly flipped). Then, split them into two lists (black, gray); 2) Reevaluate the bits in the black list, flip them according to the evaluation. Then, recalculate the syndrome; 3) Reevaluate the bits in the gray list, and flip according to the evaluation. Then, recalculate the syndrome.

The Backflip decoder has the following steps: 1) Perform 1 iteration of the Simple-Parallel decoder. For every flipped bit assign a value k. This value indicates that if the algorithm does not end after k iterations, this bit should be flipped back; 2) Flip some bits back according to their k values.

The Backflip$^+$ is variant of Backflip that uses a fixed number iterations as explained in Sect. 1. Technically, the difference is that the condition on the weight of s is moved from the `while` loop to the `if` statement (line 10). This performs the appropriate number of mock iterations.

Remark 1. The decoders use the term iterations differently. For example, one iteration of the Black-Gray decoder is somewhat equivalent to three iterations of the Simple-Parallel decoder. The iteration of the One-Round decoder consists of multiple (not necessarily fixed) "internal" iterations. Comparison of the decoders needs to take this information into account. For example, the performance is determined by the number of iterations times the latency of an iteration, not just by the number of iterations.

3 Idealized Schemes and Concrete Instantiations

We discuss some subtleties related to the requirements from a concrete algorithm in order to be acceptable as substitute for an ideal primitive, and the relation to a concrete implementation.

Cryptographic schemes are often analyzed in a framework where some of the components are modeled as ideal primitives. An ideal primitive is a black-box algorithm that performs a defined flow over some (secret) input and communicates the resulting output (and nothing more). A concrete instantiation of the scheme is the result of substituting the ideal primitive(s) with some specific algorithm(s). We require the following property from the instantiation to consider it *acceptable*: the algorithm should be *possible* to implement without communicating more information than the expected output. From the practical viewpoint, this implies that the algorithm *could be* implemented in constant-time. Note that a specific implementation of an acceptable instantiation of a provably secure scheme can still be insecure (e.g., due to side channel leakage). Special care is needed for algorithms that run with a variable number of steps.

Remark 2. A scheme can have provable security but this does not imply that every instantiation inherits the security properties guaranteed by the proof, or that there even exists an instantiation that inherits them, and an insecure instantiation example does not invalidate the proof of the idealized scheme. For example, an idealized KEM can have an IND-CCA secure proof when using a "random oracle" ideal primitive. An instantiation that replaces the random oracle with a non-cryptographic hash function does not inherit the security proof, but it is commonly acceptable to believe that an instantiation with SHA256 does.

Algorithms with a Variable Number of Steps. Let $\mathcal{A}$ be an algorithm that takes a secret input in and executes a flow with a variable number of steps/iterations $v(in)$ that depends on in. It is not necessarily possible to implement $\mathcal{A}$ in constant-time. In case ("limited") that there is a public parameter b such that $v(in) \leq b$ we can define an equivalent algorithm ($\mathcal{A}^+$) that runs in exactly b iterations: $\mathcal{A}^+$ executes the $v(in)$ iterations of $\mathcal{A}$ and continues with some $b - v(in)$ identical mock iterations. With this definition, we can assume that it is possible to implement $\mathcal{A}^+$ in constant-time. Clearly, details must be provided, and such an implementation needs to be worked out. This could be a challenging task.

Suppose that $v(in)$ is unlimited, i. e., there is no (a-priori) parameter b such that $v(in) \leq b$ (we call this case "unlimited"). It is possible to set a constant parameter b^* and an algorithm $\mathcal{A}^+$ with exactly b^* iterations, such that it emits a failure indication if the output is not obtained after exhausting the b^* iterations. It is possible to implement $\mathcal{A}^+$ in constant-time, but it is no longer equivalent to $\mathcal{A}$, due to the nonzero failure probability. Thus, analysis of $\mathcal{A}^+$ needs to include the dependency of the failure probability on b^*, and consider the resulting implications. Practical considerations would seek the smallest b^* for which the upper bound on the failure probability is satisfactory. Obviously, if $\mathcal{A}$ has originally some nonzero failure probability, then $\mathcal{A}^+$ has a larger failure probability.

Suppose that a cryptographic scheme relies on an ideal primitive. In the limited case an instantiation that substitutes $\mathcal{A}$ (or $\mathcal{A}^+$) is acceptable. However, in the unlimited case, substituting the primitive $\mathcal{A}$ with $\mathcal{A}^+$ is more delicate, due to the failure probability that is either introduced or increased. We summarize the unlimited case as follows.

- To consider $\mathcal{A}$ as an acceptable ideal primitive substitute, $v(in)$ needs to be considered as part of its output, and the security proof should take this information into consideration. Equivalently, the incremental advantage that an adversary can gain from learning $v(in)$ needs to be added to the adversary advantage of the (original) proof.
- Considering $\mathcal{A}^+$ as an acceptable ideal primitive substitute, requires a proof that it has all the properties of the ideal primitive used in the original proof (in particular, the overall failure probability).

Example 1. Consider the IND-CPA RSA PKE. Its model proof relies on the existence of the ideal primitive MODEXP for the decryption (MODEXP (a, x, N) = $a^x \pmod N$, where x is secret). Suppose that a concrete instantiation substitutes MODEXP with the square-and-multiply algorithm (*S&M*). *S&M* has a variable number of steps, $t = bitlength(x) + wt(x)$ (modular multiplications), that depends on (secret) x, where $wt(t) \leq bitlength(x)$ is the Hamming weight of x. By definition, in RSA PKE we have that $x < \phi(N) < N$ so x is a-priori bounded and consequently the number of steps in *S&M* is bounded by $t \leq 2 \cdot bitlength(x) < 2 \cdot bitlength(N)$. It is easy to define an equivalent algorithm ($S\&M^+$) that runs in exactly $2 \cdot bitlength(N)$ steps by adding mock operations. An instantiation that substitutes *S&M* (through $S\&M^+$) for MODEXP can therefore be considered acceptable (up to the understanding of how to define mock steps). This is independent of the practical security of an implementation of RSA PKE instantiated with $S\&M^+$. Such an implementation needs to account for various (side-channel) leaks e.g., branches and memory access patterns. These considerations are attributed to the implementation rather than to the instantiation, because we can trust the possibility to build such a safe implementation.

Application to BIKE. The IND-CCA security proof of BIKE relies on the existence of a decoder primitive that has a negligible DFR. This is a critical decoder's property that is used in the proof. The concrete BIKE instantiation substitutes the idealized decoder with the Backflip decoding algorithm. Backflip has the required negligible DFR. By its definition, Backflip runs in a variable number of steps (iterations) that depends on the input and on the secret key (this property is built into the algorithm's definition).

It is possible to use Backflip in order to define Backflip$^+$ decoder that has a fixed number of steps: a) Fix a number of iterations as a parameter X_{BF}; b) Follow the original Backflip flow but always execute X_{BF} iterations in a way that: if the errors vector (e) is extracted after $Y < X_{BF}$ iterations, execute additional $(X_{BF} - Y)$ identical mock iterations that do not change e; c) After the X_{BF} iterations are exhausted, output a success/failure indication and e on success or a random vector of the expected length otherwise. The difficulty is that the DFR of Backflip$^+$ is a function of X_{BF} (and r) and it may be larger from the DFR of Backflip that is critical for the proof.

It is not clear from [3,20] whether the Backflip decoder is an example of the limited or the unlimited case, but we choose to assume the limited case, based on the following indications. Backflip is defined in [3, Algorithm 4] and the definition is followed by the comment: "The algorithm takes as input [...] and, if it stops, returns an error [...] with high probability, the algorithm stops and returns e". This comment suggests the unlimited case. Here, it is difficult to accept it as a substitution of the ideal primitive, and claim that the IND-CCA security proof applies to this instantiation. In order to make Backflip an ideal primitive substitute, the number of executed steps needs to be considered as part of its output as well. As an analogy, consider a KEM where the decapsulation has nonzero failure probability. Here, an IND-CCA security proof cannot simply rely on the (original) Fujisaki-Okamoto transformation [9], because this would model an ideal decapsulation with no failures. Instead, it is possible to use the $FO^{\not\perp}$ transformation suggested in [12] that accounts for failures. This is equivalent to saying that the modeled decapsulation outputs a shared key and a success/fail indication. Indeed, this transformation was used in the BIKE CCA proof.

On the other hand, we find locations in [3], that state: "In all variants of BIKE, we will consider the decoding as a black box running in bounded time" (Sect. 2.4.1 of [3]) and "In addition, we will bound the running time (as a function of the block size r) and stop with a failure when this bound is exceeded" (Sect. 1.3 of [3]). No bounds and dependency on r are provided. However, if we inspect the reference code [3], we can find that the code sets a *maximal* number of Backflip iterations to 100 (no explanation for this number is provided and this constant is independent of r). Therefore, we may choose to interpret the results of [3,20] as if the 2^{-128} DFR was obtained from simulations with this $X_{BF} = 100$ bound[5], although this is nowhere stated and the simulation data and the derivation of the DFR are also not provided (the reference code operates with $X_{BF} = 100$). With this, it is reasonable to hope that if we take Backflip$^+$ and set $X_{BF} = 100$ we would get a DFR below 2^{-128} and this makes BacklFlip with $X_{BF} = 100$ an acceptable instantiation of an IND-CCA secure BIKE (for the studied r values).

[5] See discussion with some extrapolation methodologies in Appendix C.

The challenge with this interpretation is that the instantiation (Backflip$^+$ and $X_{BF} = 100$) would be impractical from the performance viewpoint. Our paper solves this by showing acceptable instantiations with a much smaller values of X_{BF}. Furthermore, it also shows that there are decoders with a fixed number of iterations that have better performance at the same DFR level.

Implementation. In order to be used in practice, an IND-CCA KEM should have a proper *instantiation* and *also* a constant-time *implementation* that is secure against side-channel attacks (e.g., [8]). Such attacks were demonstrated in the context of QC-MDPC schemes, e.g., the GJS reaction attack [11] and several subsequent attacks [8,15,19]. Other reaction attacks examples include [18] for LRPC codes and [22] for attacking the repetition code used by the HQC KEM [2]. This problem is significantly aggravated when the KEM is used with static keys (e.g., [5,8]).

4 Estimating the DFR of a Decoder with a Fixed Number of Iterations

The IND-CCA BIKE proof assumes a decapsulation algorithm that invokes an ideal decoding primitive. Here, the necessary condition is that the decapsulation has a negligible DFR, e.g., 2^{-128} [3,12]. Therefore, a technique to estimate the DFR of a decoder is an essential tool.

The Extrapolation Method of [20]**.** An extrapolation method technique for estimating the DFR is shown in [20]. It consists of the following steps: a) Simulate proper encapsulation and decapsulation of random inputs for small block sizes (r values), where a sufficiently large number of failures can be observed; b) Extrapolate the observed data points to estimate the DFR for larger r values.

The DFR analyses in [20] and [3] applies this methodology to decoders that have some maximum number of iterations X_{BF} (we choose to assume that $X_{BF} = 100$ was used). In our experiments Backflip$^+$ always succeeds/fails before reaching 100 iterations for the relevant values of r. Practically, it means that setting $X_{BF} = 100$ can be considered equivalent to setting an unlimited number of iterations.

Our goal is to estimate the DFR of a decoder that is allowed to perform exactly X iterations (where X is predefined). We start from small values of X (e.g., $X = 2, 3, \ldots$) and increase it until we no longer see failures (in a large number of experiments) caused by exhausting X iterations. Larger values of X lead to a smaller DFR.

We tested BIKE-1 and BIKE-3 in Level-1 and Level-3 with the Black-Gray and the Backflip$^+$ decoders. In order to acquire a sufficient number of data points we tested multiple prime r values such that $x^r - 1$ is a primitive polynomial [3]. The specific values are listed in Appendix D.

For our study, we used a slightly different extrapolation method. For every combination (scheme, level, decoder, r) we ran $N_{exp} = 48,000,000$ experiments as follows: a) Generate, uniformly at random, a secret key and an errors vector (e), compute the public key, and perform encapsulation-followed-by-decapsulation (with e); b) Allow the decoder to run up to $X = 100$ iterations[6]; c) Record the actual number of iterations that were required in order to recover e. If the decoder exhausts the 100 iterations it stops and marks the experiment as a decoding failure. For every $X < 100$ we say that the X-DFR is the sum of the number of experiments that fail (after 100 iterations) plus the number of experiments that required more than X iterations divided by N_{exp}. Next, we fix the scheme, the level, the decoder, and X, and we end up with an X-DFR value for every tested r. Subsequently, we perform linear/quadratic extrapolation on the data and receive a curve. We use this curve to find the value r_0 for which the X-DFR is our target probability p_0 and use the pair (r_0, p_0) as the BIKE scheme parameters.

We target three p_0 values: a) $p_0 = 2^{-23} \approx 10^{-7}$ that is reasonable for most practical use cases (with IND-CPA schemes); b) $p_0 = 2^{-64}$ also for an IND-CPA scheme but with a much lower DFR; c) $p_0 = 2^{-128}$, which is required for an IND-CCA Level-1 scheme. The linear/quadratic functions and the resulting r_0 values are given in Appendix E [Table 4].

Our Extrapolation Methodology. In most cases, we were able to confirm the claim of [20] that the evolution of the DFR as a function of r occurs in two phases: quadratic initially, and then linear. As in [20], we are interested in extrapolating the linear part because it gives a more conservative DFR approximation. We point out that the results are sensitive to the method used for extrapolation (see details in Appendix C). Therefore, it is important to define it precisely so that the results can be reproduced and verified. To this end, we determine the starting point of the linear evolution as follows: going over the different starting points, computing the fitting line and picking the one for which we get the best fit to the data points. Here, the merit of the experimental fit is measured by the L2 norm (i. e., mean squared error). The L2 norm is a good choice in our case, where we believe that the data may have a few outliers.

5 Results

A description of the Backflip$^+$ constant-time implementation is provided in Appendix B.

The Experimentation Platform. Our experiments were executed on an AWS EC2 `m5.metal` instance with the 6th Intel® Core^{TM} Generation (Micro Architecture Codename "Sky Lake" [SKL]) Xeon® Platinum 8175M CPU 2.50 GHz. It has 384 GB RAM, 32K L1d and L1i cache, 1MiB L2 cache, and 32MiB L3 cache, where the Intel® Turbo Boost Technology was turned off.

The Code. The core functionality was written in x86−64 assembly and wrapped by assisting C code. The code uses the `PCLMULQDQ`, `AES-NI` and the `AVX2` and

[6] Recall that different decoders have different definition for the term "iterations", see Sect. 2.3.

AVX512 instructions. The code was compiled with gcc (version 7.4.0) in 64-bit mode, using the "O3" Optimization level, and run on a Linux (Ubuntu 18.04.2 LTS) OS. It uses the NTL library [21] compiled with the GF2X library [17].

Figure 8 in Appendix F shows the simulation results for BIKE-1, Level-1 and Level-3, using the Black-Gray and Backflip$^+$ decoders. Note that we use the IND-CCA flows. The left panels present linear extrapolations and the right panels present quadratic extrapolations. The horizontal axis measures the block size r in bits, and the vertical axis shows the simulated $log_{10}(DFR)$ values. Every panel displays several graphs associated with different X values. The minimal X is chosen so that the extrapolated r value for $DFR = 2^{-128}$ is still considered to be secure according to [3]. The maximal value of X is chosen to allow a meaningful extrapolation. We give two examples:

Example 2. Consider Black-Gray. Typically, there exists some number of iterations $j < X_{BG}$, where if decoding a syndrome requires more than j then the decoder fails (w.h.p) even if a large number of iterations X_{BG} is allowed.

The quadratic approximations shown in Fig. 8 yield a nice fit to the data points. However, we prefer to use the more pessimistic linear extrapolation in order to determine the target r.

Validating the Extrapolation. We validated the extrapolated results for every extrapolation graph. We chose some r that is not a data point on the graph (but is sufficiently small to allow direct simulations). We applied the extrapolation to obtain an estimated DFR value. Then, we ran the simulation for this value of r and compared the results. Table 2 shows this comparison for several values of r and the Black-Gray decoder with $X_{BG} = 3$. We note that for 10,267 and 10,301 we tested at least 960 million and 4.8 billion tests respectively. In case of 10,301 decoding always succeeded after $X_{BG} = 4$ iterations, while for 10,267 there were too few failures for meaningful computation of the DFR. Therefore, we use $X_{BG} = 3$ in our experimentation in order to observe enough failures. For example, the extrapolation for the setting (BIKE-1, Level-1, Black-Gray, 10,301) estimates 3-DFR $= 10^{-7.55}$ this is very close to the experimented DFRs $10^{-7.56}$.

Table 2. Validating the extrapolation results for the Black-Gray decoder with $X_{BG} = 3$ over two values of r.

r	Extrapolated DFR	Experimented DFR	Number of tests
10,267	$10^{-7.13}$	$10^{-7.26}$	9.6e8
10,301	$10^{-7.55}$	$10^{-7.56}$	4.8e9

5.1 Extensive Experimentation

To observe that the Black-Gray decoder does not fail in practice with $r = 11,779$ (i. e., the recommended r for the Backflip decoder) we run extensive simulations. We executed $10^{10} \approx 2^{33}$ tests that generate a random key, encapsulate a message and decapsulate the resulting ciphertext. Indeed, we did not observe any decoding failure (as expected).

5.2 Performance Studies

The performance measurements reported hereafter are measured in processor cycles (per single core), where lower count is better. All the results were obtained using the same measurement methodology, as follows. Each measured function was isolated, run 25 times (warm-up), followed by 100 iterations that were clocked (using the `RDTSC` instruction) and averaged. To minimize the effect of background tasks running on the system, every experiment was repeated 10 times, and the minimum result was recorded.

For every decoder, the performance depends on: a) X - the number of iterations; b) the latency of one iteration. Recall that comparing *just* the number of iterations is meaningless. Table 3 provides the latency ($\ell_{decoder,r}$) of one iteration and the overall decoding latency ($l_{decoder,r,i} = X_{decoder} \cdot \ell_{decoder,r}$) for the Black-Gray and the Backflip^{+} decoders, for several values of r. The first four rows of the table report for the value $r = 10,163$ that corresponds to the BIKE-1-CPA proposal, and for the value $r = 11,779$ that corresponds to the BIKE-1-CCA proposal. The following rows report values of r for which the decoders achieve the same DFR.

Clearly, the constant-time Black-Gray decoder is faster than the constant-time Backflip^{+} decoder (when both are restricted to a given number of iterations).

We now compare the performance of the BIKE-1-CCA flows to the performance of the BIKE-1-CPA flows, for given r values, using the Black-Gray decoder with $X_{BG} = 3, 4$. Note that values of r that lead to DFR $> 2^{-128}$ cannot give IND-CCA security. Furthermore, even with BIKE-1-CCA flows and r such that DFR $\leq 2^{-128}$, IND-CCA security is not guaranteed (see the discussion in Sect. 6). The results are shown in Fig. 2. The bars show the total latency of the key generation (blue), encapsulation (orange), and decapsulation (green) operations. The slowdown imposed by using the BIKE-1-CCA flows compared to using the BIKE-1-CPA flows is indicated (in percents) in the figure. We see that the additional cost of using BIKE-1-CCA flows is only ~6% in the worst case.

6 Weak Keys: A Gap for Claiming IND-CCA Security

Our analysis of the decoders, the new parameters suggestion, and the constant-time implementation makes significant progress towards a concrete instantiation

Table 3. A performance comparison of the Black-Gray and the Backflip$^+$ decoders for BIKE-1 Level-1. The r values were chosen according to Table 4.

DFR	Decoder	r	$X_{decoder}$	$\ell_{decoder,r}$ (cycles)	$l_{decoder,r,i}$ (million cycles)
2^{-19}	Black-Gray	10,163	3	702,785	2.1
2^{-17}	Backflip$^+$	10,163	8	751,246	6.0
2^{-101}	Black-Gray	11,779	4	784,903	3.13
2^{-58}	Backflip$^+$	11,779	9	841,806	6.73
2^{-23}	Black-Gray	10,253	3	743,168	2.22
2^{-23}	Black-Gray	10,163	4	702,785	2.8
2^{-23}	Backflip$^+$	10,499	8	777,478	6.22
2^{-23}	Backflip$^+$	10,253	9	764,959	6.88
2^{-64}	Black-Gray	11,261	3	769,212	2.3
2^{-64}	Black-Gray	11,003	4	769,820	3.0
2^{-64}	Backflip$^+$	12,781	8	907,905	7.26
2^{-64}	Backflip$^+$	12,011	9	856,084	7.7
2^{-128}	Black-Gray	12,781	3	849,182	2.54
2^{-128}	Black-Gray	12,347	4	841,310	3.36
2^{-128}	Backflip$^+$	14,797	9	1,024,798	9.22

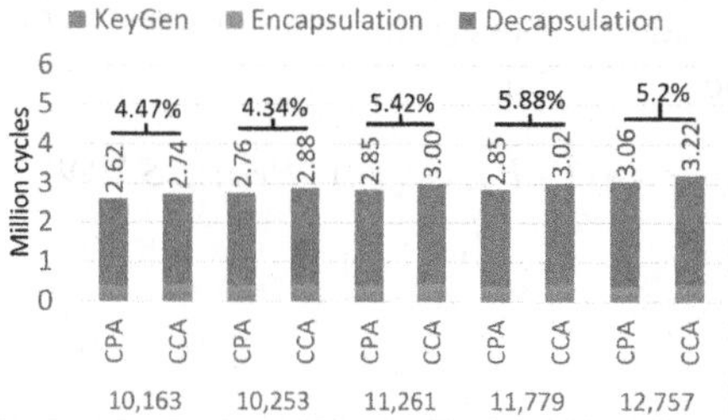

(a) $X_{BG} = 3$

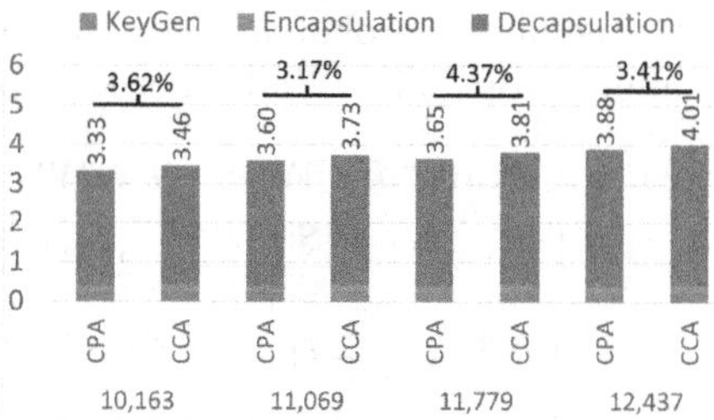

(b) $X_{BG} = 4$

Fig. 2. Comparison of BIKE-1-CPA flows and BIKE-1-CCA flows, running with the Black-Gray decoder and $X_{BG} = 3, 4$ for several values of r: $r = 10,163$ the original BIKE-1-CPA; $r = 11,779$ the original BIKE-1-CCA; r values that correspond to DFR of $2^{-23}, 2^{-64}, 2^{-128}$, according to Table 4. Note that values of r that lead to $DFR > 2^{-128}$ do not give IND-CCA security. The vertical axis measures latency in millions of cycles (lower is better). The additional cost of using the IND-CCA flows it at most 6%.

and implementation of IND-CCA BIKE. However, we believe that there is still a subtle missing gap for claiming IND-CCA security, that needs to be addressed.

The remaining challenge is that a claim for IND-CCA security depends on having an underlying δ-correct PKE (for example with $\delta = 2^{-128}$ for Level-1) [12]. This notion is different from having a DFR of 2^{-128}, and leads to the following problem. The specification [3] defines the DFR as "the probability for the decoder to fail when the input (h_0, h_1, e_0, e_1) is distributed uniformly". The δ-correctness property of a PKE/KEM is defined through Eqs. (1), (2) above. These equations imply that δ is the average of the *maximum* failure probability taken over all the possible messages. By contrast, the DFR notion relates to the average probability.

Remark 3. We also suggest to fix a small inaccuracy in the statement of the BIKE-1 proof [3]: "... the resulting KEM will have the exact same DFR as the underlying cryptosystem ...". Theorem 3.1 of [12] states that: "If PKE is δ-correct, then PKE_1 is $\delta 1$-correct in the random oracle model with $\delta 1(qG) = qG \cdot \delta.$[...]". Theorem 3.4 therein states that: "If PKE_1 is $\delta 1$-correct then $\text{KEM}^{\not\perp}$ is $\delta 1$-correct in the random oracle model [..]"[7]. Thus, even if DFR $= \delta$, the statement should be "the resulting KEM is $(\delta \cdot qG)$-correct, where the underlying PKE is δ-correct".

To illustrate the gap between the definitions, we give an example for what can go wrong.

Example 3. Let $\mathcal{S}$ be the set of valid secret keys and let $|\mathcal{S}|$ be its size. Assume that a group of weak keys $\mathcal{W}$ exists[8], and that $\frac{|\mathcal{W}|}{|\mathcal{S}|} = \bar{\delta} > 2^{-128}$. Suppose that for every key in $\mathcal{W}$ there exists at least one message for which the probability in Eq. (1) equals 1. Then, we get that $\delta > \bar{\delta} > 2^{-128}$. By comparison the average DFR can still be upper bounded by 2^{-128}. For instance, let $|\mathcal{S}| = 2^{130}$, $|\mathcal{W}| = 2^4$ and let the failure probability over all messages for every weak key be 2^{-10}. Let the failure probability of all other keys be 2^{-129}. Then,

$$\begin{aligned}
DFR &= Pr(\texttt{fail} \mid k \in \mathcal{W}) \cdot Pr(k \in \mathcal{W}) + Pr(\texttt{fail} \mid k \in \mathcal{S} \setminus \mathcal{W}) \cdot Pr(k \in \mathcal{S} \setminus \mathcal{W}) \\
&= \frac{|\mathcal{W}|}{|\mathcal{S}|} \cdot 2^{-10} + \frac{|\mathcal{S}| - |\mathcal{W}|}{|\mathcal{S}|} \cdot 2^{-129} \\
&= 2^{-126} \cdot 2^{-10} + (1 - 2^{-126}) \cdot 2^{-129} \\
&= 2^{-136} + 2^{-129} - 2^{-255} < 2^{-128}
\end{aligned}$$

6.1 Constructing Weak Keys

Currently, we are not aware of studies that classify weak keys for QC-MDPC codes or bound their numbers. To see why this gap cannot be ignored we designed a series of tests that show the existence of a relatively large set of weak keys. Our examples are inspired by the notion of "spectrum" used in [8,11,15]. To construct the keys we changed the BIKE key generation. Instead of generating a random h_0, we start by setting the first $f = 0, 20, 30, 40$ bits, and then select randomly the positions of the additional $(d - f)$ bits. The generation of h_1 is unchanged.

[7] Here, $\text{KEM}^{\not\perp}$ refers to a KEM with implicit rejection, and qG is the number of invocation of the random oracle G (H in the case of BIKE-1).

[8] Our definition of weak keys is different form that of [4], where a weak key is a secret key that can be exposed from the public key alone.

Since it is difficult to observe failures and weak keys behavior when r is large, we study $r = 10,163$ (of BIKE-1 CPA) and also $r = 9,803$ that amplify the phenomena.

Figure 7 in Appendix F shows the behavior of the Black-Gray decoder for $r = 9,803$ and $r = 10,163$ with $f = 0, 20, 30, 40$ after $X_{BG} = 1, 2, 3, 4$ iterations. In every case (Panel) we choose randomly 10,000 keys. For every key we choose randomly 1,000 errors vectors. The histograms show the weight of an "ideal" errors vector e after the X_{BG} iteration (horizontal axis). We see that as f grows the number of un-decoded error bits after every iteration increases. For $f \leq 30$, decoding often succeeds after 3 iterations. However, for $f = 40$ it is possible to decode the error after 4 iterations only when $r = 10,163$, but not for $r = 9,803$. In other words, if we fix $X_{BG} = 3$ for the Black-Gray decoder and use $r = 9,803$ we see ~100% failures with weak keys defined by $f = 40$. This shows that for a given decoder the set of weak keys depends on r and X.

Remark 4 (Other decoders). Figure 7 shows how the weak keys impact the decoding success probability for chosen r and X_{BG} with the Black-Gray decoder. Note that such results depend on the specific decoder. To compare, Backflip$^+$ calculates the unsatisfied parity checks threshold in a more conservative way, and therefore requires more iterations. Weak keys would lead to a different behavior. When we repeat our experiment with the Simple-Parallel decoder, we see that almost all tests fail even with $f = 19$.

Figure 3 shows additional results with the Black-Gray decoder and $r = 9,803$. Panel (a) shows the histogram for $f = 0$ (i.e., reducing to the standard h_0 generation), and Panel (b) shows the histogram for $f = 30$. The horizontal axis measures the number of failures x out of 10,000 random errors. The vertical axis counts the number of keys that have a DFR of $x/10,000$. For $f = 0$, the average and standard deviation are $\mathbb{E}(x) = 119.06$, and $\sigma(x) = 10.91$. However, when $f = 30$, the decoder fails much more often and we have $\mathbb{E}(x) = 9,900.14$, and $\sigma(x) = 40.68$. This shows the difference between the weak keys and the "normal" randomized keys and that the DFR cannot be predicted by the "average-case" model. It is also interesting to note that for $f = 30$ we do not get a Gaussian like distribution (unlike the histogram with $f = 0$).

The remaining question is: what is the probability to hit a weak key when the keys are generated randomly as required? Let $\mathcal{W}_f$ be the set of weak keys that correspond to a given value of f. Define $z_{r,f}$ as the relative size of $\mathcal{W}_f$. Then

$$z_{r,f} = \frac{|\mathcal{W}_f|}{|\mathcal{S}|} = \frac{\binom{r-f}{d-f}}{\binom{r}{d}} \tag{3}$$

Note that choosing a larger f decreases the size of $\mathcal{W}_f$, i.e., if $f_2 < f_1$ then $\mathcal{W}_{f_1} \subseteq \mathcal{W}_{f_2}$. It is easy to compute that

$$z_{9803,0} = z_{10163,0} = 1$$

$$z_{9803,10} = 2^{-72}, \quad z_{9803,20} = 2^{-146}, \quad z_{9803,30} = 2^{-223}, \quad z_{9803,40} = 2^{-304},$$

$$z_{10163,10} = 2^{-72}, \quad z_{10163,20} = 2^{-147}, \quad z_{10163,30} = 2^{-225}, \quad z_{10163,40} = 2^{-306}$$

The conclusion is that while the set $\mathcal{W}_f$ is large, its relative size (from the set of all keys) is still below 2^{-128}. Therefore, this construction *does not* show that BIKE-1 after our fix is necessarily *not* IND-CCA secure. However, it clearly shows that the problem cannot be ignored, and the claim that BIKE *is* IND-CCA secure requires further justification. In fact, there are other patterns and combinations that give sets of weak keys with a higher relative size (e.g., setting every other bit of h_0, f times). We point out again that any analysis of weak keys should relate to a *specific* decoder and a *specific* choice of r.

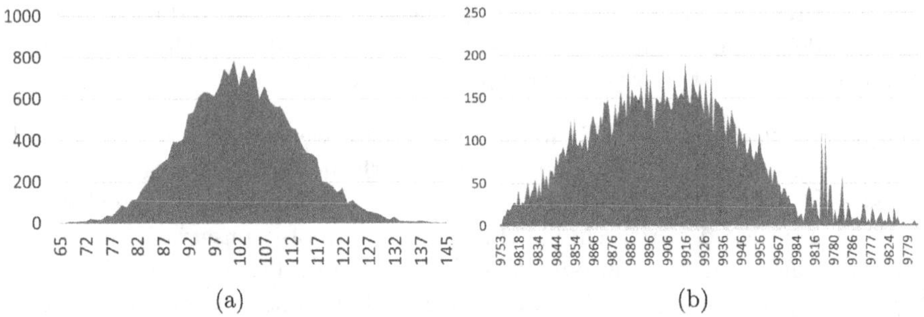

Fig. 3. Black-Gray decoder, $r = 9,803$ with $f = 0$ (Panel (a)) and $f = 30$ (Panel (b)). The horizontal-axis measures the number of failures x out of 10,000 random errors vectors. The vertical-axis counts the number of keys that have a DFR of $x/10,000$. The conclusion is that there are keys that lead to higher DFR.

7 Discussion

The Round-2 BIKE [3] represents significant progress in the scheme's design, and offers an IND-CCA version, on top of the IND-CPA KEM that was defined in Round-1. This paper addresses several difficulties and challenges and solves some of them.

- The Backflip decoder runs in a variable number of steps that depends on the input and the secret key. We fix this problem by defining a variant, Backflip$^+$, that, by definition, runs X_{BF} iterations for a parameter X_{BF}. We carry out the analysis to determine the values of X_{BF} where Backflip$^+$ has DFR of 2^{-128}, and provide all of the details that are needed in order to repeat and reproduce our experiments. Furthermore, we show that for the target DFR, the values of X_{BF} are relatively small e.g., 12. (much less than 100 as implied for Backflip).
- Inspired by the extrapolation method suggested in [20], we studied the Black-Gray decoder (already used in Round-1 Additional code [7]) that we defined to have a fixed number of steps (iterations) X_{BG}. Our goal was to find values of X_{BG} that guarantee the target DFR for a given r. We found that the values

of r required with Black-Gray are smaller than the values with Backflip$^+$. It seems that achieving the low DFR (2^{-128}) should be attributed to increasing r, independently of the decoding algorithm. The ability to prove this for some decoders is attributed to the extrapolation method.

- After the decoders are defined to run a fixed number of iterations, we could build constant-time software implementations (with memory access patterns and branches that reveal no secret information). This is nowadays the standard expectation from cryptographic implementations. We measured the resulting performance (on a modern x86-64 CPU) to find an optimal "decoder-X-r" combination. Table 3 shows that for a given DFR, the Black-Gray decoder is always faster than Backflip$^+$.
- The analysis in Sect. 6 identifies a gap that needs to be addressed in order to claim IND-CCA security for BIKE. It relates to the difference between average DFR and the δ-correctness definition [12]. A DFR of (at most) 2^{-128} is a necessary requirement for IND-CCA security, which BIKE achieves. However, it is not necessarily sufficient. We show how to construct a "large" set of weak keys, but also show that it still not sufficiently large to invalidate the necessary δ-correctness bound. This is a positive indication, although there is no proof that the property is indeed satisfied. This gap remains as an interesting research challenge to pursuit. The problem of bounding (or eliminating) the number of weak keys is not specific to BIKE. It is relevant for other schemes that claim IND-CCA security and their decapsulation has nonzero failure probability. With this, we can state the following.

> BIKE-1-CCA, instantiated with Black-Gray (or Backflip$^+$) decoder with the parameters that guarantee DFR of 2^{-128}, and with the accompanying constant-time implementation, is IND-CCA secure, under the assumption that its underlying PKE is 2^{-128}-correct.

7.1 Methodologies

Our performance measurements were carried on an x86-64 architecture. Studies on different architectures can give different results we therefore point to an interesting study of the performance of other constant-time decoders on other platforms [14]. Note that [14] targets schemes that use ephemeral keys with relatively large DFR and only IND-CPA security.

Differences in the DFR Estimations. Our DFR prediction methodology may be (too) conservative and therefore yields more pessimistic results than those of [3]. One example is the combination (BIKE-1, Level-1, Backflip$^+$ decoder, $r = 11,779$, $X_{BF} = 10$). Here, [3] predicts a 100−DFR of 2^{-128} and our linear extrapolation for the 10-DFR predicts only 2^{-71}($\approx 10^{-21}$). To achieve a 10-DFR of 2^{-128} we need to set $r = 13,892$(>11,779). Although the Backflip$^+$ decoder with $X_{BF} = 10$ is not optimal, it is important to understand the effect of different extrapolations. Comparing to [3,20] is difficult (no information that allows us to repeat the experiments), we attempt to provide some insight by acquiring data points for Backflip$^+$ with $X_{BF} = 100$ and applying our extrapolation methodology. Indeed, the results we obtain are still more pessimistic, but if we apply a different extrapolation methodology ("Last-Linear") we get closer to [3]. The details are given in Appendix C.

Another potential source of differences is that Backflip has a recovery mechanism (TTL). For Backflip$^+$ this mechanism is limited due to setting $X_{BF} \leq 11$. It may be possible to tune Backflip and Backflip$^+$ further by using some fine-grained TTL equations that depend on r. Information on the equations that were used for [3] was not published, so we leave tuning for further research.

7.2 Practical Considerations for BIKE[9]

Our Decoder Choice. We report our measurements only for Black-Gray and Backflip$^+$ because other decoders that we know either have a worse DFR (e.g., Parallel-Simple) or are inherently slow (e.g., Step-by-Step). Our results suggest that instantiating BIKE with Black-Gray is recommended. We note that the higher number of iterations required by Backflip$^+$ is probably because it uses a more conservative threshold function than Black-Gray.

Recommendations for the BIKE Flows. Currently, BIKE has two options for executing the key generation, encapsulation, and decapsulation flows. One for an IND-CPA KEM, and another (using the $FO^{\not\perp}$ transformation [12]) for an IND-CCA scheme, to deny a chosen ciphertext attack from the encapsulating party. It turns out that the performance difference is relatively small. As shown in Fig. 2 for BIKE-1, the overhead of the IND-CCA flows is less than 6% (on x86-64 platforms). With such a low overhead, we believe that the BIKE proposal could gain a great deal of simplification by using *only* the IND-CCA flows. This is true even for applications that intend to use only ephemeral keys, in order to achieve forward secrecy. Here, IND-CCA security is not mandatory, and IND-CPA security suffice. However, using the $FO^{\not\perp}$ transformation could be a way to reduce the risk of inadvertent repetition ("misuse") of a supposedly ephemeral key, thus buying some multi-user-multi-key robustness. By applying this approach, the scheme is completely controlled by choosing a single parameter r (with the same implementation). For example, with the Black-Gray decoder and $X_{BG} = 4$, the choice $r = 11,001$ gives DFR $= 2^{-64}$ with competitive performance. A DFR of 2^{-64} is sufficiently low and can be ignored by all practical considerations.

Choosing r. The choice of r and X_{BG} gives an interesting trade-off between bandwidth and performance[10]. A larger value of r increases the bandwidth but reduces the DFR when X_{BG} is fixed. On the other hand, it allows to reduce

[9] The recommendations given here are the opinion of the authors of this paper.

[10] BIKE specification [3, Section 2.4.5] states: "An interesting consequence is that if w and t are fixed, a moderate modification of r (say plus or minus 50%) will not significantly affect the resistance against the best known key and message attacks".

X_{BG} while maintaining the same DFR. This could lead to better performance. We give one example. To achieve DFR $= 2^{-23}$ the choice of $X_{BG} = 4$ and $r = 10,163$ leads to decoding at $2.8M$ cycles. The choice $X_{BG} = 3$ and a slightly larger $r = 10,253$ leads to decoding at $2.22M$ cycles. Complete details are given in Table 3.

General Recommendations for the BIKE Suite. Currently, BIKE [3] consists of 10 variants: BIKE-1 (the simplest and fastest); BIKE-2 (offering bandwidth optimization at the high cost of polynomial inversion); BIKE-3 (simpler security reduction with the highest bandwidth and lowest performance). In addition, there are also BIKE-2-batch and BIKE-3 with bandwidth optimization. Every version comes with two flavors IND-CPA and IND-CCA. On top of this, every option comes with three security levels (L1/L3/L5). Finally, the implementation packages include generic code and optimization for AVX2 and AVX512. We believe that this abundance of options involves too high complexity and reducing their number would be very useful. For Round-3 we recommend to define BIKE as follows: BIKE-1 CCA instantiated with the Black-Gray decoder with $X_{BG} = 3$ iterations. Offer Level-1 with $r = 11,261$ targeting DFR $= 2^{-64}$ and $r = 12,781$ targeting DFR $= 2^{-128}$, as the main variants. In all cases, use ephemeral keys, for forward secrecy. For completeness, offer also a secondary variant for Level-3 with $r = 24,659$ targeting DFR $= 2^{-128}$.

The code that implements these recommendations was contributed to (and already merged into) the open-source library LibOQS [1]. It uses the choice of $r = 11,779$, following the block size of the current Round-2 specification (this choice of r leads to a DFR of 2^{-86}).

Vetting Keys. We recommend to use BIKE with ephemeral keys and forward secrecy. In this case we do not need to rely on the full IND-CCA security properties of the KEM. However, there may be usages that prefer to use static keys. Here, we recommend the following way to narrow the DFR-δ-correctness gap pointed above by "vetting" the private key. For static keys we can assume that the overall latency of the key generation generation phase is less significant. Therefore, after generating a key, it would be still acceptable, from the practical viewpoint, to vet it experimentally. This can be done by running encapsulation-followed-by-decapsulation for some number of trials, in the hope to identify a case where the key is obviously weak. A more efficient way is to generate random (and predefined) errors and invoke the decoder. We point out that the vetting process can also be applied offline.

Acknowledgments. This research was partly supported by: NSF-BSF Grant 2018640; The BIU Center for Research in Applied Cryptography and Cyber Security, and the Center for Cyber Law and Policy at the University of Haifa, both in conjunction with the Israel National Cyber Directorate in the Prime Minister's Office.

A   Black-Gray Decoder

Algorithm 2. e=Black-Gray(c, H)

Input: Parity-check matrix $H \in \mathbb{F}_2^{r \times n}$, $c \in \mathbb{F}_2^n$, X_{BG} (maximal # of iterations)
Output: The error $e \in \mathbb{F}_2^n$
Exception: "decoding failure" return $\perp$
1: **procedure** BLACK-GRAY(c, H)
2:     $s = Hc^T$, $e = 0$, $\delta = 4$
3:     $B = \emptyset$, $G = \emptyset$ ▷ Black and Gray position sets
4:
5:     **for** itr in $0 \ldots X_{BG} - 1$ **do**
6:         $th = $ computeThreshold(s)
7:         $upc[n - 1 : 0] = computeUPC(s, H)$
8:         **for** i in $0 \ldots n - 1$ **do**   ▷ Step I
9:             **if** $upc[i] \geq th$ **then**
10:                 $e[i] = e[i] \oplus 1$ ▷ Flip an error bit
11:                 $B = B \cup i$ ▷ Update the Black set
12:             **else if** $upc[i] > th - \delta$ **then**
13:                 $G = G \cup i$ ▷ Update the Gray set
14:         $s = H(c^T + e^T)$ ▷ Update the syndrome
15:
16:         $upc[n - 1 : 0] = computeUPC(s, H)$ ▷ Step II
17:         **for** $b \in B$ **do**
18:             **if** $upc[b] > ((d + 1)/2)$ **then**
19:                 $e[b] = e[b] \oplus 1$ ▷ Flip an error bit
20:         $s = H(c^T + e^T)$ ▷ Update the syndrome
21:
22:         $upc[n - 1 : 0] = computeUPC(s, H)$ ▷ Step III
23:         **for** $g \in G$ **do**
24:             **if** $upc[g] > ((d + 1)/2)$ **then**
25:                 $e[g] = e[g] \oplus 1$ ▷ Flip an error bit
26:         $s = H(c^T + e^T)$ ▷ Update the syndrome
27:
28:     **if** $(wt(s) \neq 0)$ **then**
29:         **return** $\perp$
30:     **else**
31:         **return** e

B   Implementing Backflip$^+$ in Constant-Time

Here, we show how to define and implement a constant-time Backflip$^+$ decoder, based on a constant-time Black-Gray decoder. The Backflip$^+$ decoder differs from the Black-Gray decoder in two aspects: a) it uses a new mechanism called TTL; b) it uses new equations for calculating the thresholds. The TTL mechanism is a "smart queue" where the decoder flips back some error bits when it believes that they were mistakenly flipped in previous iterations. It does so unconditionally and it can flip bits even after 5 iterations. The Black-Gray decoder uses a different type of TTL, where the black and gray lists serve as the "smart queue". However, the error bits are flipped back after only 1 iteration, conditionally, through checking certain thresholds. Indeed, as we report below the differences are observed in cases where the Black-Gray decoder failed to decode after 4 iterations and then w.h.p fails completely. The Backflip decoder shows better recovery capabilities in such cases. Implementing the new TTL queue in constant-time relies mostly on common constant-time techniques.

Handling the New Threshold Function. The Backflip decoder thresholds are a function of two variables [3][Section 2.4.3]: a) the syndrome weight $wt(s)$ as in the Black-Gray decoder; b) the number of error bits that the decoder believes it flipped (denoted $\bar{e}$). This function outputs higher thresholds compared to the Black-Gray decoder. This is a conservative approximation. We believe that the design of the Backflip decoder tends to avoid flipping the "wrong" bits so that the decoder would have better recovery capabilities and a lower DFR (assuming that it can execute an un-bounded number of iterations). We point out that evaluating the function involves computing logarithms, exponents, and function minimization, and it is not clear how this can be implemented in constant-time (the reference code [3] is not implemented in constant-time).

One way to address this issue is to pre-calculate the finite number of pairs $(wt(s), \bar{e})$ and their function evaluation, store them in a table, and read them from the table in constant-time. This involves very high latencies.

Similarly to the Black-Gray decoder (in BIKE-1-CPA [3]), we approximate the thresholds function - which is here a function of two variables. A first attempt is shown in Fig. 4. We compute the function over all the valid/relevant inputs and then compute an approximation by fitting it to a plane. Unfortunately, this approximation is not sufficiently accurate, an experiment with $r = 11,779$ (as in BIKE-1-CCA [3]) gave an estimated DFR of 10^{-4} .

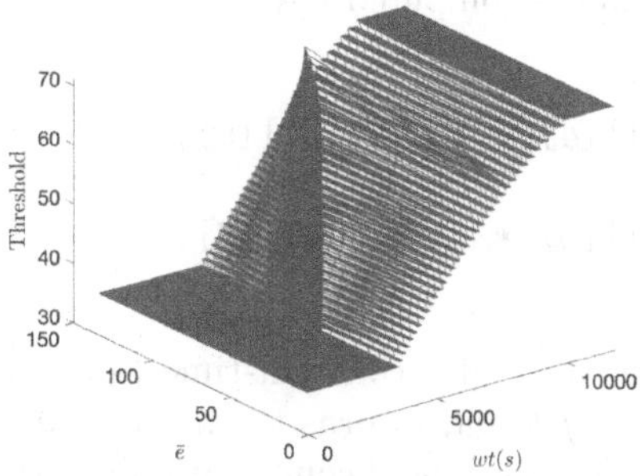

(a) The thresholds function.

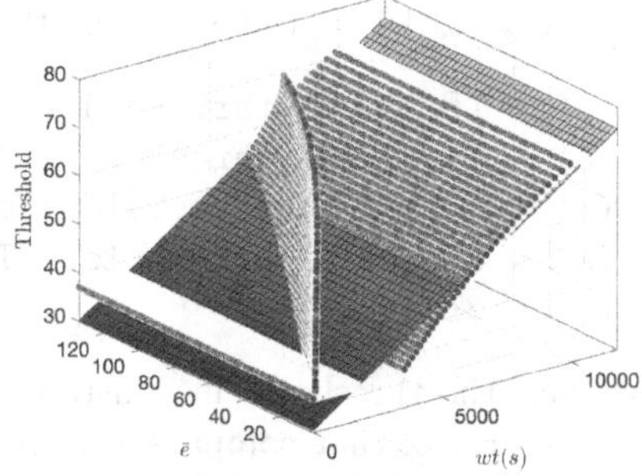

(b) Approximating the function (blue) using a plane (yellow,turquoise).

Fig. 4. Approximating the Backflip decoder thresholds function.

To improve the approximation we project the function onto the plane $\bar{e} = e1$ $(0 \leq e1 \leq t)$. Then, for every valid $e1$, we compute the linear approximation and tabulate the coefficients. Figure 5, Panel (a) illustrates the linear approximation for $e1 = 25$. These thresholds improve the DFR but it is still too high.

A refinement can be obtained by partitioning the approximation into five regions. The projection graph in Fig. 5 can be partitioned in five intervals as follows: a) $[a_0, a_1]$, $[a_2, a_3]$, where the threshold is fixed to some minimum value (min); b) $[a_4, a_5]$ where the threshold is d; c) $[a_1, a_2]$ and $[a_3, a_4]$ where the threshold (th) is approximated using $th = b_0 wt(s) + b_1$ and $th = c_0 wt(s) + c_1$, respectively. For $r = 11,779$ the values we use are $a_0 = 0$, $a_1 = 1,578$, $a_2 = 1,832$ $a_3 = 3,526$, $a_4 = 9,910$ $a_5 = r$. The results is shown Fig. 5 Panel (b) for $\bar{e} = 25$. We use these values to define the table (T) with t rows and 8 columns. Every row contains the a_1, a_2, a_3, a_4, b_0, b_1, c_0, c_1 values that correspond to the projection on the plane $\bar{e}$.

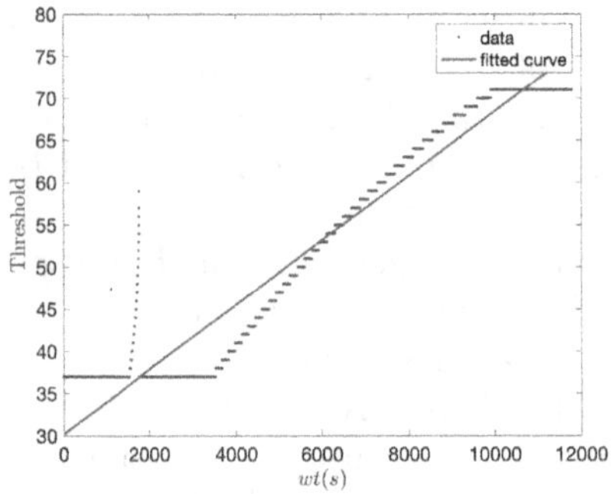

(a) Linear approximation of $wt(s) \in [0, r]$

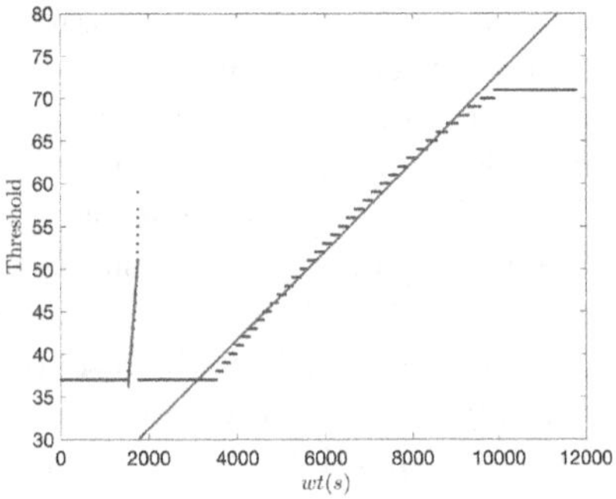

(b) Five linear approximation per part.

Fig. 5. Approximating the threshold function when $\bar{e} = 25$ is fixed.

For every $(s1, e1) = (wt(s), \bar{e})$ the threshold is computed by

```
if   (s1 < T[e1][0]) threshold = min;
elif (s1 < T[e1][1]) threshold = T[e1][4] * s1 + T[e1][5];
elif (s1 < T[e1][2]) threshold = min;
elif (s1 < T[e1][3]) threshold = T[e1][6] * s1 + T[e1][7];
else threshold = d
```

To evaluate the thresholds in constant-time we used a constant-time function `secure_le_mask` that compares two integers j, k and returns the mask `0x0` if $j < k$ and the mask `0xffffffff` otherwise. The threshold computation is now:

```
cond0 = secure_le_mask(T[e1][0], s1)
cond1 = secure_le_mask(T[e1][1], s1) & ~secure_le_mask(T[e1][0], s1)
cond2 = secure_le_mask(T[e1][2], s1) & ~secure_le_mask(T[e1][1], s1)
cond3 = secure_le_mask(T[e1][3], s1) & ~secure_le_mask(T[e1][2], s1)
cond4 = ~secure_le_mask(T[e1][3],s1)

res  = cond0 & min
res += cond1 & round(T[e1][4] * s1 + T[e1][5])
res += cond2 & min
res += cond3 & round(T[e1][6] * s1 + T[e1][7])
res += cond4 & max
return res
```

With this we can implement Backflip$^+$ in constant-time, provided that we fix a-priori the number of iterations.

C Achieving the Same DFR Bounds as of [20]

We ran experiments with Backflip$^+$ and $X_{BF} = 100$ for BIKE-1 Level-1, scanning all the 34 legitimate $r \in [8500, 9340]$ (prime r values such that $x^r - 1$ is a primitive polynomial) with $4.8M$ tests for every value. Applying our extrapolation methodology (see Sect. 4) to the acquired data leads to the results illustrated in Fig. 6 Panels (a) and (b). The figure highlights the pairs (DFR; r) for DFR 2^{-64} and 2^{-128} with the smallest possible r. For example, with $r = 12,539$ the linear extrapolation gives DFR of 2^{-128}. Note that [3] claims a DFR of 2^{-128} for a smaller $r = 11,779$. For comparison, with $r = 11,779$ our methodology gives a DFR of 2^{-104}. We can guess that either different TTL values were used for every r, or that other r values were used, or that a different extrapolation methodology was applied.

We show one possible methodology ("Last-Linear") that gives a DFR of $\sim 2^{-128}$ with $r = 11,779$ when applied to the acquired data: a) Ignore the points from the data-set for which 100−DFR is too low to be calculated reliably (e.g., the five lower points in Fig. 6); b) Draw a line through the last two remaining data points with the highest values of r. The rationale is that the "linear regime" of the DFR evolution starts for values of r that are beyond those that can be estimated in an experiment. Thus, a line drawn through two data points where r is smaller than the starting point of the linear regime leads to an extrapolation that is lower-bounded by the "real" linear evolution. With this approach, the question is how to choose the two points for which experimental data is obtained and the DFR is extrapolated from.

This shows that different ways to acquire and interpret the data give different upper bounds for the DFR. Since the extrapolation shoots over a large gap of r values, the results are sensitive to the chosen methodology. It is interesting to note that if we take our data points for Black-Gray and $X_{BG} = 5$ and use the Last-Linear extrapolation, we can find two points that would lead to 2^{-128} and $r = 11,779$, while more conservative methodology gives only 2^{-101}.

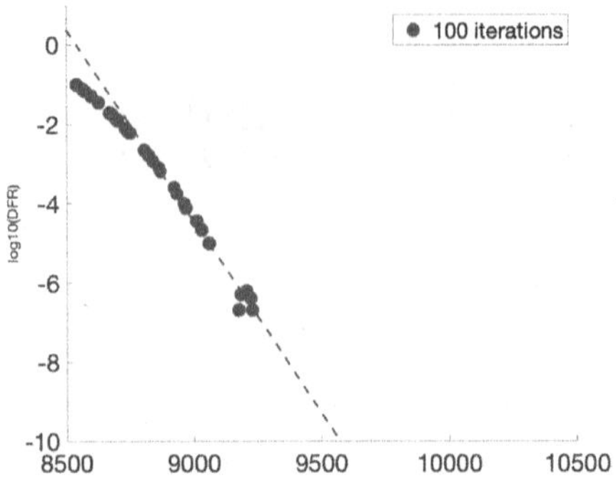

(a) lin. ext., our method. (DFR, r) = (2^{-64}; 10, 589), (2^{-128}; 12, 539)

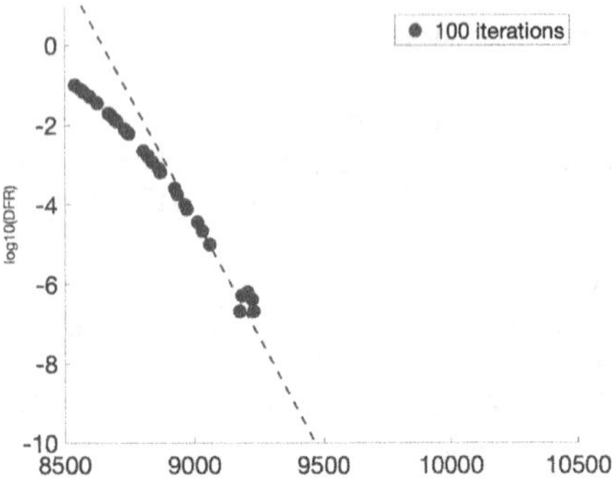

(b) lin. ext. Last-Linear. (DFR, r) = (2^{-64}; 10, 253), (2^{-128}; 11, 813)

Fig. 6. BIKE-1 Level-1 Backflip$^+$ different extrapolation methods. See the text for details. The sub-captions detail the (DFR; r) for DFR values: 2^{-64}, 2^{-128}.

D Additional Information

The following values of r were used by the extrapolation method:

- BIKE-1 Level-1: 9349, 9547, 9749, 9803, 9859, 9883, 9901, 9907, 9923, 9941, 9949, 10037, 10067, 10069, 10091, 10093, 10099, 10133, 10139.
- BIKE-1 Level-3: 19013, 19037, 19051, 19069, 19141, 19157, 19163, 19181, 19219, 19237, 19259, 19301, 19333, 19373, 19387, 19403, 19427, 19469, 19483, 19507, 19541, 19571, 19597, 19603, 19661, 19709, 19717, 19739, 19763, 19813, 19853.

E The Linear and the Quadratic Extrapolations

Table 4 gives the equations for the linear and the quadratic extrapolation together with the extrapolated values of r for a DFR of 2^{-23}, 2^{-64}, and 2^{-128}. It covers the tuple (scheme, level, decoder, X), where decoder $\in$ {BG=Black-Gray, BF=Backflip$^+$}.

The BIKE specification [3] chooses r to be the minimum required for achieving a certain security level, and the best bandwidth trade-off. It also indicates that it is possible to increase r by "plus or minus 50%" (leaving w, t fixed) without reducing the complexity of the best known key/message attacks. This is an interesting observation. For example, increasing the BIKE-1 Level-3 $r = 19,853$ by 50% gives $r = 29,779$ which is already close to the BIKE-1 Level-5 that has $r = 32,749$ (of course with different w and t). We take a more conservative approach and restrict r values to be at most 30% above their CCA values stated in [3]. Table 4 labels values beyond this limit as N/A.

Table 4. The linear and the quadratic extrapolation equations, and the computed r values for a given DFR. The cases labeled with N/A are those where the value of r to achieve a target DFR could not be found in the range $[0.7r', 1.3r']$, where r' is the recommended value for IND-CCA security in [3]

KEM	Lev.	Decoder	Iter.	Lin. start	Lin. eq. (a,b) s.t. $\log_{10}$ DFR $= ar + b =$	2^{-23}	2^{-64}	2^{-128}	Quad. eq. (a, b, c) s.t. $\log_{10}$ DFR $= ar^2 + br + c =$	2^{-23}	2^{-64}	2^{-128}
BIKE-1	1	BG	3	10	(−1.25e−2, 121)	10,253	11,261	12,781	(−1.05e−5, 1.97e−1, −927)	10,253	10,789	11,317
BIKE-1	1	BG	4	9	(−1.45e−2, 140)	10,163	11,003	12,347	(−1.16e−5, 2.18e−1, −1020)	10,139	10,667	11,197
BIKE-1	1	BG	5	9	(−1.49e−2, 144)	10,133	10,973	12,251	(−1.18e−5, 2.20e−1, −1030)	10,133	10,667	11,171
BIKE-1	1	BF	8	9	(−5.40e−3, 49.8)	10,499	12,781	N/A	(−6.86e−7, 8.11e−3, −16.8)	10,459	12,149	14,107
BIKE-1	1	BF	9	6	(−6.92e−3, 63.8)	10,253	12,011	14,797	(−1.16e−6, 1.62e−2, −50.9)	10,253	11,579	13,109
BIKE-1	1	BF	10	8	(−8.40e−3, 77.6)	10,067	11,549	13,829	(−1.88e−6, 2.90e−2, −108)	10,067	11,197	12,437
BIKE-1	1	BF	11	7	(−1.12e−2, 104)	9,949	11,069	12,781	(−3.41e−6, 5.77e−2, −243)	9,949	10,883	11,867
BIKE-1	3	BG	3	10	(−6.97e−3, 133)	20,051	21,821	24,659	(−2.39e−6, 8.64e−2, −780)	19,997	21,059	22,189
BIKE-1	3	BG	4	10	(−8.70e−3, 166)	19,853	21,269	23,459	(−3.34e−6, 1.22e−1, −1110)	19,813	20,717	21,683
BIKE-1	3	BG	5	10	(−9.10e−3, 173)	19,813	21,139	23,251	(−3.67e−6, 1.34e−1, −1220)	19,763	20,627	21,557
BIKE-1	3	BF	8	10	(−5.36e−3, 99.7)	19,867	22,171	25,771	(−9.08e−7, 3.02e−2, −248)	19,853	21,523	23,339
BIKE-1	3	BF	9	9	(−6.14e−3, 114)	19,661	21,661	24,781	(−1.37e−6, 4.71e−2, −403)	19,661	21,059	22,613
BIKE-1	3	BF	10	5	(−6.51e−3, 120)	19,469	21,379	24,371	(−8.64e−7, 2.69e−2, −204)	19,469	21,011	22,787
BIKE-1	3	BF	11	6	(−7.05e−3, 130)	19,373	21,101	23,869	(−1.66e−6, 5.69e−2, −488)	19,373	20,693	22,067

F Illustration Graphs

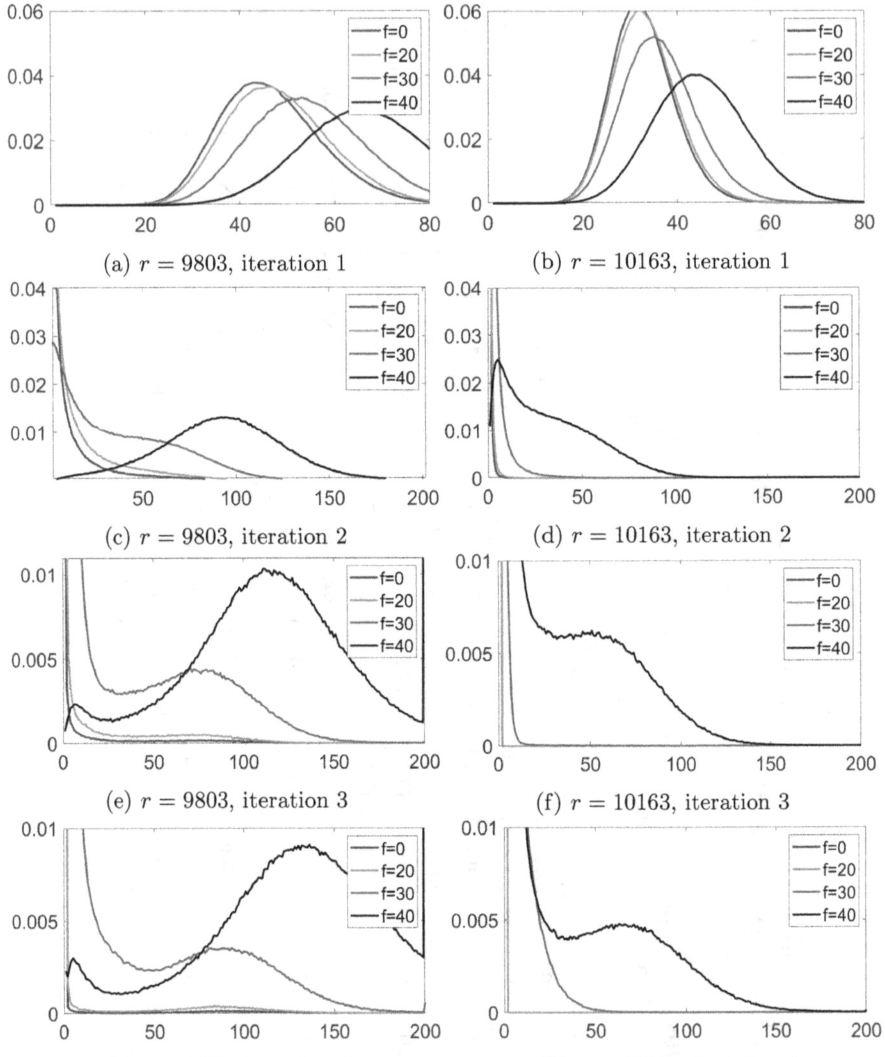

(a) $r = 9803$, iteration 1

(b) $r = 10163$, iteration 1

(c) $r = 9803$, iteration 2

(d) $r = 10163$, iteration 2

(e) $r = 9803$, iteration 3

(f) $r = 10163$, iteration 3

(g) $r = 9803$, iteration 4

(h) $r = 10163$, iteration 4

Fig. 7. Histograms of the cases (vertical axis; measured in percentage) that end-up with some weight of an "ideal" errors vector (horizontal axis) after the $X_{BG} = 1, 2, 3, 4$ iterations. The decoder is the Black-Gray decoder. Panels a, c, e, g represents the results for $r = 9,803$ and Panels b, d, f, h for $r = 10,163$ with $f = 0, 20, 30, 40$. A lower error weight is better. See explanation in the text.

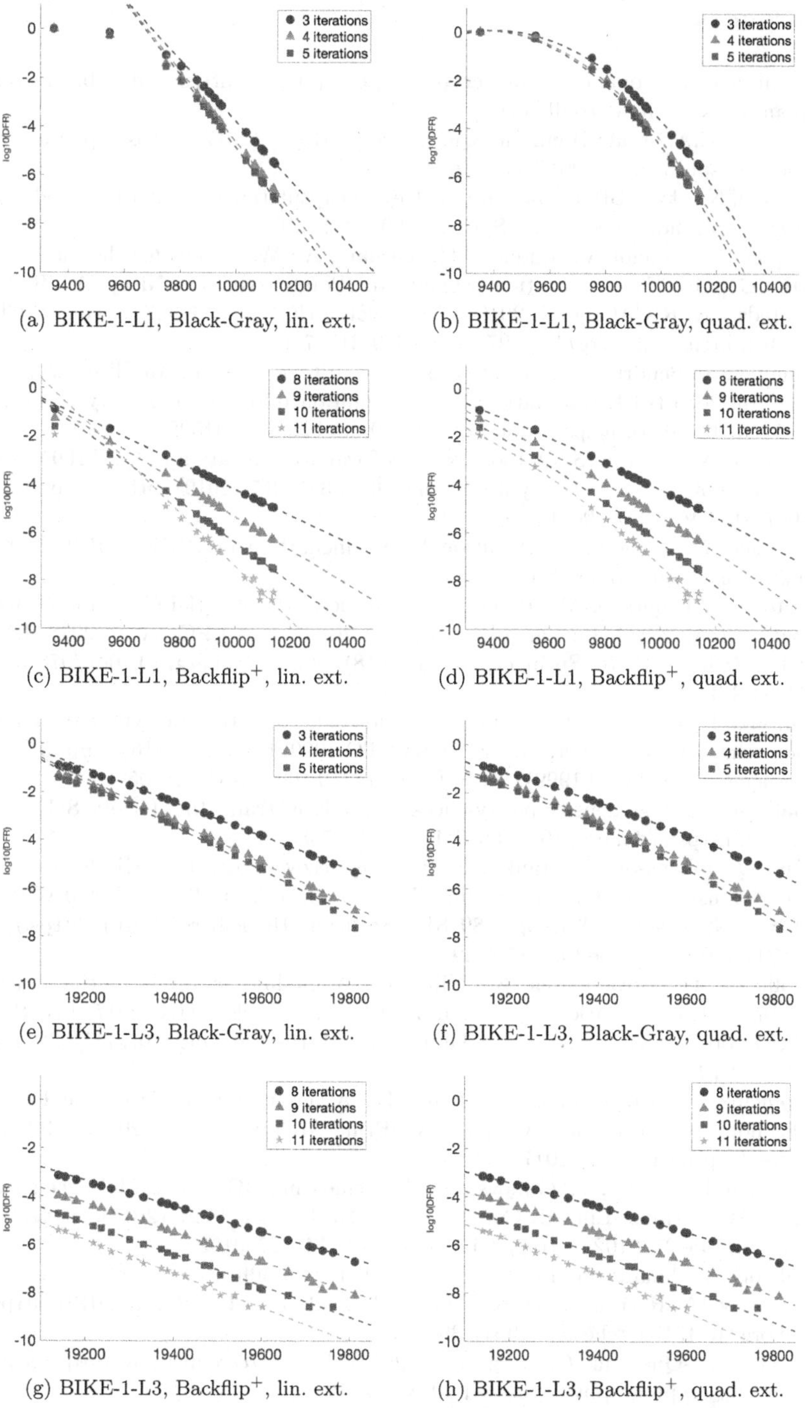

(a) BIKE-1-L1, Black-Gray, lin. ext.

(b) BIKE-1-L1, Black-Gray, quad. ext.

(c) BIKE-1-L1, Backflip$^+$, lin. ext.

(d) BIKE-1-L1, Backflip$^+$, quad. ext.

(e) BIKE-1-L3, Black-Gray, lin. ext.

(f) BIKE-1-L3, Black-Gray, quad. ext.

(g) BIKE-1-L3, Backflip$^+$, lin. ext.

(h) BIKE-1-L3, Backflip$^+$, quad. ext.

Fig. 8. BIKE-1 Level-1 and Level-3 extrapolations (see Sect. 4 for details).

References

1. C library for quantum-safe cryptography (2019). https://github.com/open-quantum-safe/liboqs/pull/554
2. Melchor, C.A., et al.: Hamming Quasi-Cyclic (HQC) (2017). https://pqc-hqc.org/doc/hqc-specification_2017-11-30.pdf
3. Aragon, N., et al.: BIKE: Bit Flipping Key Encapsulation (2017). https://bikesuite.org/files/round2/spec/BIKE-Spec-2019.06.30.1.pdf
4. Bardet, M., Dragoi, V., Luque, J.G., Otmani, A.: Weak keys for the quasi-cyclic MDPC public key encryption scheme. In: Pointcheval, D., Nitaj, A., Rachidi, T. (eds.) AFRICACRYPT 2016. LNCS, vol. 9646, pp. 346–367. Springer, Cham (2016). https://doi.org/10.1007/978-3-319-31517-1_18
5. Chaulet, J., Sendrier, N.: Worst case QC-MDPC decoder for McEliece cryptosystem. In: 2016 IEEE International Symposium on Information Theory (ISIT), pp. 1366–1370 (2016). https://doi.org/10.1109/ISIT.2016.7541522
6. Drucker, N., Gueron, S.: A toolbox for software optimization of QC-MDPC code-based cryptosystems. J. Cryptogr. Eng. **9**(4), 341–357 (2019). https://doi.org/10.1007/s13389-018-00200-4
7. Drucker, N., Gueron, S.: Additional implementation of BIKE (2019). https://bikesuite.org/additional.html
8. Eaton, E., Lequesne, M., Parent, A., Sendrier, N.: QC-MDPC: a timing attack and a CCA2 KEM. In: Lange, T., Steinwandt, R. (eds.) PQCrypto 2018. LNCS, vol. 10786, pp. 47–76. Springer, Cham (2018). https://doi.org/10.1007/978-3-319-79063-3_3
9. Fujisaki, E., Okamoto, T.: Secure integration of asymmetric and symmetric encryption schemes. In: Wiener, M. (ed.) CRYPTO 1999. LNCS, vol. 1666, pp. 537–554. Springer, Heidelberg (1999). https://doi.org/10.1007/3-540-48405-1_34
10. Gallager, R.: Low-density parity-check codes. IRE Trans. Inf. Theory **8**(1), 21–28 (1962). https://doi.org/10.1109/TIT.1962.1057683
11. Guo, Q., Johansson, T., Stankovski, P.: A key recovery attack on MDPC with CCA security using decoding errors. In: Cheon, J.H., Takagi, T. (eds.) ASIACRYPT 2016. LNCS, vol. 10031, pp. 789–815. Springer, Heidelberg (2016). https://doi.org/10.1007/978-3-662-53887-6_29
12. Hofheinz, D., Hövelmanns, K., Kiltz, E.: A modular analysis of the Fujisaki-Okamoto transformation. In: Kalai, Y., Reyzin, L. (eds.) TCC 2017. LNCS, vol. 10677, pp. 341–371. Springer, Cham (2017). https://doi.org/10.1007/978-3-319-70500-2_12
13. Hofheinz, D., Hövelmanns, K., Kiltz, E.: A modular analysis of the Fujisaki-Okamoto transformation. Cryptology ePrint Archive, Report 2017/604 (2017). https://eprint.iacr.org/2017/604
14. Maurich, I.V., Oder, T., Güneysu, T.: Implementing QC-MDPC McEliece encryption. ACM Trans. Embed. Comput. Syst. **14**(3), 44:1–44:27 (2015). https://doi.org/10.1145/2700102. http://doi.acm.org/10.1145/2700102
15. Nilsson, A., Johansson, T., Wagner, P.S.: Error amplification in code-based cryptography. IACR Trans. Cryptogr. Hardw. Embed. Syst. **1**, 238–258 (2019). https://doi.org/10.13154/tches.v2019.i1.238-258
16. NIST: Post-Quantum Cryptography (2019). https://csrc.nist.gov/projects/post-quantum-cryptography. Accessed 20 Aug 2019
17. Gaudry, P., Brent, R.P.Z., Thome, E.: gf2x-1.2 (2017). https://gforge.inria.fr/projects/gf2x/

18. Samardjiska, S., Santini, P., Persichetti, E., Banegas, G.: A reaction attack against cryptosystems based on LRPC codes. In: Schwabe, P., Thériault, N. (eds.) LATINCRYPT 2019. LNCS, vol. 11774, pp. 197–216. Springer, Cham (2019). https://doi.org/10.1007/978-3-030-30530-7_10
19. Santini, P., Battaglioni, M., Chiaraluce, F., Baldi, M.: Analysis of reaction and timing attacks against cryptosystems based on sparse parity-check codes. In: Baldi, M., Persichetti, E., Santini, P. (eds.) CBC 2019. LNCS, vol. 11666, pp. 115–136. Springer, Cham (2019). https://doi.org/10.1007/978-3-030-25922-8_7
20. Sendrier, N., Vasseur, V.: On the decoding failure rate of QC-MDPC bit-flipping decoders. In: Ding, J., Steinwandt, R. (eds.) PQCrypto 2019. LNCS, vol. 11505, pp. 404–416. Springer, Cham (2019). https://doi.org/10.1007/978-3-030-25510-7_22
21. Shoup, V.: Number theory C++ library (NTL) version 11.3.4 (2019). http://www.shoup.net/ntl
22. Wafo-Tapa, G., Bettaieb, S., Bidoux, L., Gaborit, P.: A practicable timing attack against HQC and its countermeasure. Technical report 2019/909 (2019). https://eprint.iacr.org/2019/909

Protograph-Based Decoding of Low-Density Parity-Check Codes with Hamming Weight Amplifiers

Hannes Bartz[1(✉)], Emna Ben Yacoub[2(✉)], Lorenza Bertarelli[3(✉)], and Gianluigi Liva[1(✉)]

[1] Institute of Communication and Navigation, German Aerospace Center (DLR), Wessling, Germany
{hannes.bartz,gianluigi.liva}@dlr.de

[2] Institute for Communications Engineering, Technical University of Munich, Munich, Germany
emna.ben-yacoub@tum.de

[3] JMA Wireless, Bologna, Italy
lbertarelli@jmawireless.com

Abstract. A new protograph-based framework for message passing (MP) decoding of low density parity-check (LDPC) codes with Hamming weight amplifiers (HWAs), which are used e.g. in the NIST post-quantum crypto candidate LEDAcrypt, is proposed. The scheme exploits the correlations in the error patterns introduced by the HWA using a turbo-like decoding approach where messages between the decoders for the outer code given by the HWA and the inner LDPC code are exchanged. Decoding thresholds for the proposed scheme are computed using density evolution (DE) analysis for belief propagation (BP) and ternary message passing (TMP) decoding and compared to existing decoding approaches. The proposed scheme improves upon the basic approach of decoding LDPC code from the amplified error and has a similar performance as decoding the corresponding moderate-density parity-check (MDPC) code but with a significantly lower computational complexity.

Keywords: Mceliece cryptosystem · LDPC codes · Hamming weight amplifiers · Code-based cryptography

1 Introduction

In 1978, McEliece proposed a code-based public-key cryptosystem (PKC) [1] that relies on the hardness of decoding an unknown linear error-correcting code. Unlike the widely-used Rivest-Shamir-Adleman (RSA) cryptosystem [2], the McEliece cryptosystem is resilient against attacks performed on a quantum computer and is considered as *post-quantum* secure. One drawback of the McEliece cryptosystem compared to the RSA cryptosystem is the large key size and the rate-loss. Many variants of the McEliece cryptosystem based on different code

M. Baldi et al. (Eds.): CBCrypto 2020, LNCS 12087, pp. 80–93, 2020.
https://doi.org/10.1007/978-3-030-54074-6_5

families have been considered. In particular, McEliece cryptosystems based on low density parity-check (LDPC) codes allow for very small keys but suffer from feasible attacks on the low-weight dual code [3].

A variant based on quasi cyclic (QC)-LDPC codes that uses a sparse column scrambling matrix, a so-called Hamming weight amplifier (HWA), to increase the density of the public code parity-check matrix was proposed in [4]. However, poor choices of the column scrambling matrix allow for structural attacks [5]. In [6] a scheme that defeats the attack in [5] by using dense row scrambling matrices and less structured column scrambling matrices was presented. Optimized code constructions for the cryptosystem proposed in [6] were presented in [7]. The ideas and results of [6,7] are the basis for the LEDAcrypt [8] PKC and authentication schemes that are candidates for the current post-quantum cryptosystem standardization by NIST.

LEDAcrypt [8] uses a variant of the bit-flipping[1] decoder [10], called "Q-decoder", that exploits the correlation in the error patterns due to the HWA. The "Q-decoder" has the same error-correction performance as a BF decoder for the corresponding moderate-density parity-check (MDPC) code but with a significantly lower computational complexity [7]. In this paper, a new protograph-based decoding scheme for LDPC codes with HWAs is presented [8, Chap. 5]. The new scheme provides a turbo-like decoding framework, where information between the decoder of an outer rate-one code given by the HWA and the decoder of the inner LDPC codes, is exchanged. The proposed framework allows to compare, analyze and optimize message passing (MP) decoding schemes for LDPC codes with HWAs.

The density evolution (DE) analysis for belief propagation (BP) and ternary message passing (TMP) decoding shows that the proposed protograph-based scheme has a similar error-correction capability as the corresponding MDPC code under MP decoding and improves upon the basic approach of decoding the amplified error using an LDPC decoder. For some parameters, the protograph-based scheme improves upon the corresponding MDPC decoding approach while having a lower computational complexity due to the sparsity of the extended graph. The gains in the error-correction capability predicted by DE analysis are validated by Monte Carlo simulations under BP and TMP decoding.

2 Preliminaries

2.1 Circulant Matrices

Denote the binary field by $\mathbb{F}_2$ and the set of $m \times n$ matrices over $\mathbb{F}_2$ by $\mathbb{F}_2^{m\times n}$. The set of all vectors of length n over $\mathbb{F}_2$ is denoted by $\mathbb{F}_2^n$. Vectors and matrices are written as bold lower-case and upper-case letters such as $\boldsymbol{a}$ and $\boldsymbol{A}$, respectively. A binary circulant matrix $\boldsymbol{A}$ of size p is a $p \times p$ matrix with coefficients in $\mathbb{F}_2$

[1] In [8] and other literature the bit-flipping (BF) decoder is referred to as "Gallager's BF" algorithm although it is different from the algorithm proposed by Gallager in [9].

obtained by cyclically shifting its first row $\boldsymbol{a} = (a_0, a_1, \ldots, a_{p-1})$ to the right, yielding

$$\boldsymbol{A} = \begin{pmatrix} a_0 & a_1 & \cdots & a_{p-1} \\ a_{p-1} & a_0 & \cdots & a_{p-2} \\ \vdots & \vdots & \ddots & \vdots \\ a_1 & a_2 & \cdots & a_0 \end{pmatrix}.$$

The set of $p \times p$ circulant matrices together with matrix multiplication and addition forms a commutative ring that is isomorphic to the polynomial ring $(\mathbb{F}_2[X]/(X^p - 1), +, \cdot)$. In particular, there is a bijective mapping between a circulant matrix $\boldsymbol{A}$ and a polynomial $a(X) = a_0 + a_1 X + \ldots + a_{p-1} X^{p-1} \in \mathbb{F}_2[X]$. We indicate the vector of coefficients of a polynomial $a(X)$ as $\boldsymbol{a} = (a_0, a_1, \ldots, a_{p-1})$. The weight of a polynomial $a(X)$ is the number of its non-zero coefficients, i.e., it is the Hamming weight of its coefficient vector $\boldsymbol{a}$. We indicate both weights with the operator $\mathsf{wht}(\cdot)$, i.e., $\mathsf{wht}(a(X)) = \mathsf{wht}(\boldsymbol{a})$. In the remainder of this paper we use the polynomial representation of circulant matrices to provide an efficient description of the structure of the codes.

2.2 McEliece Cryptosystem Using LDPC Codes with Hamming Weight Amplifiers

For $n = N_0 p$, dimension $k = K_0 p$, redundancy $r = n - k = R_0 p$ with $R_0 = N_0 - K_0$ for some integer p, a parity-check matrix $\boldsymbol{H}(X)$ of a QC-LDPC[2] code in polynomial form is a $R_0 \times N_0$ matrix where each entry (polynomial) describes the corresponding circulant matrix. We denote the corresponding $R_0 \times N_0$ base matrix that indicates the Hamming weights of the polynomials in $\boldsymbol{H}(X)$ by

$$\boldsymbol{B}_H = \begin{pmatrix} b_{00} & b_{01} & \cdots & b_{0(N_0-1)} \\ b_{10} & b_{11} & \cdots & b_{1(N_0-1)} \\ \vdots & \vdots & \ddots & \vdots \\ b_{(R_0-1)0} & b_{(R_0-1)1} & \cdots & b_{(R_0-1)(N_0-1)} \end{pmatrix}.$$

The column scrambling matrix $\boldsymbol{Q}(X)$ is of the form

$$\boldsymbol{Q}(X) = \begin{pmatrix} q_{00}(X) & \cdots & q_{0(N_0-1)}(X) \\ \vdots & \ddots & \vdots \\ q_{(N_0-1)0}(X) & \cdots & q_{(N_0-1)(N_0-1)}(X) \end{pmatrix}. \tag{1}$$

[2] As in most of the literature, we loosely define a code to be QC if there exists a permutation of its coordinates such that the resulting (equivalent) code has the following property: if $\boldsymbol{x}$ is a codeword, then any cyclic shift of $\boldsymbol{x}$ by ℓ positions is a codeword. For example, a code admitting a parity-check matrix as an array of $R_0 \times N_0$ circulants does not fulfill this property. However the code is QC in the loose sense, since it is possible to permute its coordinates to obtain a code for which every cyclic shift of a codeword by $\ell = N_0$ positions yields another codeword.

We denote the corresponding base matrix for $\boldsymbol{Q}(X)$ by

$$\boldsymbol{B}_Q = \begin{pmatrix} b_0 & b_1 \dots b_{N_0-1} \\ b_{N_0-1} & b_0 \dots b_{N_0-2} \\ \vdots & \vdots \ddots \quad \vdots \\ b_1 & b_2 \dots \quad b_0 \end{pmatrix}$$

where $\sum_{i=0}^{N_0-1} b_i = d_Q$. This implies that $\boldsymbol{Q}(X)$ has constant row and column weight d_Q.

Without loss of generality we consider in the following codes with $r = p$ (i.e. $R_0 = 1$). This family of codes covers a wide range of code rates and is of particular interest for cryptographic applications since the parity-check matrices can be characterized in a compact way. QC-LDPC codes with $r = p$ admit a parity-check matrix of the form

$$\boldsymbol{H}(X) = \left(h_0(X)\ h_1(X) \dots h_{N_0-1}(X)\right). \tag{2}$$

Let $\text{DEC}_{\boldsymbol{H}}(\cdot)$ be an efficient decoder for the code defined by the parity-check matrix $\boldsymbol{H}$ that returns an estimate of a codeword or a decoding failure.

Key Generation

- Randomly generate a parity-check matrix $\boldsymbol{H} \in \mathbb{F}_2^{r\times n}$ of the form (2) with $\mathsf{wht}\left(h_i(X)\right) = d_c^{(i)}$ for $i = 0, \dots, N_0 - 1$ and an invertible column scrambling matrix $\boldsymbol{Q} \in \mathbb{F}_2^{n\times n}$ of the form (1). The matrix $\boldsymbol{H}$ with low row weight $d_c = \sum_{i=0}^{N_0-1} d_c^{(i)}$ and the matrix $\boldsymbol{Q}$ with low row and column weight d_Q is the *private* key.
- From the private matrices $\boldsymbol{H}(X)$ and $\boldsymbol{Q}(X)$ the matrix $\boldsymbol{H}'(X)$ is obtained as

  $$\boldsymbol{H}'(X) = \boldsymbol{H}(X)\boldsymbol{Q}(X) = \left(h'_0(X) \dots h'_{N_0-1}(X)\right).$$

  The row weight d'_c of $\boldsymbol{H}'(X)$ is upper bounded by

  $$d'_c \leq d_c d_Q.$$

  Due to the low density of $\boldsymbol{H}$ and $\boldsymbol{Q}$ we have $d'_c \approx d_c d_Q$. Hence, the density of $\boldsymbol{H}'$ is higher than $\boldsymbol{H}$ which results in a degraded error-correction performance. Depending on d_c and d_Q, $\boldsymbol{H}'$ may be a parity-check matrix of a QC-MDPC code [11].
- The *public* key is the corresponding binary $k \times n$ generator matrix $\boldsymbol{G}'(X)$ of $\boldsymbol{H}'(X)$ in systematic form[3].

[3] We assume that $G'(x)$ can be brought into systematic form which is possible with high probability (see [8]).

Encryption

- To encrypt a plaintext[4] $\boldsymbol{u} \in \mathbb{F}_2^k$ a user computes the ciphertext $\boldsymbol{c} \in \mathbb{F}_2^n$ using the public key $\boldsymbol{G}'$ as
$$\boldsymbol{c} = \boldsymbol{u}\boldsymbol{G}' + \boldsymbol{e}$$
where $\boldsymbol{e}$ is an error vector uniformly chosen from all vectors from $\mathbb{F}_2^n$ of Hamming weight $\mathsf{wht}(\boldsymbol{e}) = e$.

Decryption

- To decrypt a ciphertext $\boldsymbol{c}$ the authorized recipient uses the secret matrix $\boldsymbol{Q}$ to compute the transformed ciphertext
$$\tilde{\boldsymbol{c}} = \boldsymbol{c}\boldsymbol{Q}^T = \boldsymbol{u}\boldsymbol{G}'\boldsymbol{Q}^T + \boldsymbol{e}\boldsymbol{Q}^T. \tag{3}$$
A decoder $\mathrm{DEC}_{\boldsymbol{H}}(\cdot)$ using the secret matrix $\boldsymbol{H}$ is applied to decrypt the transformed ciphertext $\tilde{\boldsymbol{c}}$ as
$$\hat{\boldsymbol{c}} = \mathrm{DEC}_{\boldsymbol{H}}(\tilde{\boldsymbol{c}}) = \mathrm{DEC}_{\boldsymbol{H}}(\boldsymbol{u}\boldsymbol{G}'\boldsymbol{Q}^T + \boldsymbol{e}\boldsymbol{Q}^T). \tag{4}$$
- The generator matrix corresponding to $\boldsymbol{H}$ is used to recover the plaintext $\boldsymbol{u}$ from $\hat{\boldsymbol{c}}$.

2.3 Protograph Ensembles

A protograph $\mathscr{P}$ [13] is a small bipartite graph comprising a set of N_0 variable nodes (VNs) (also referred to as VN types) $\{\mathsf{V}_0, \mathsf{V}_1, \ldots, \mathsf{V}_{N_0-1}\}$ and a set of M_0 check nodes (CNs) (i.e., CN types) $\{\mathsf{C}_0, \mathsf{C}_1, \ldots, \mathsf{C}_{M_0-1}\}$. A VN type V_j is connected to a CN type C_i by b_{ij} edges. A protograph can be equivalently represented in matrix form by an $M_0 \times N_0$ matrix $\boldsymbol{B}$. The jth column of $\boldsymbol{B}$ is associated to VN type V_j and the ith row of $\boldsymbol{B}$ is associated to CN type C_i. The (i, j) element of $\boldsymbol{B}$ is b_{ij}. A larger graph (derived graph) can be obtained from a protograph by applying a copy-and-permute procedure. The protograph is copied p times (p is commonly referred to as lifting factor), and the edges of the different copies are permuted preserving the original protograph connectivity: If a type-j VN in the derived graph is connected to a type-i CN with b_{ij} edges in the protograph, each type-j VN is connected to b_{ij} distinct type-i CNs (multiple connections between a VN and a CN are not allowed in the derived graph). The derived graph is a Tanner graph with $n_0 = N_0 p$ VNs and $m_0 = M_0 p$ CNs that represent a binary linear block code. A protograph $\mathscr{P}$ defines a code ensemble $\mathscr{C}$. For a given protograph $\mathscr{P}$, consider all its possible derived graphs with $n_0 = N_0 p$ VNs. The ensemble $\mathscr{C}$ is the collection of codes associated to the derived graphs in the set.

A distinctive feature of protographs is the possibility of specifying graphs that have VNs which are associated to codeword symbols, as well as VNs which are

[4] We assume that the CCA-2 security conversions from [12] are applied to the McEliece cryptosystem to allow for systematic encoding without security reduction.

not associated to codeword symbols. The latter class of VNs are often referred to as *state* or *punctured* VNs. The term "punctured" is used since the code associated with the derived graph can be seen as a punctured version of a longer code associated with the same graph for which all the VNs are associated to codeword bits. The introduction of state VNs in a code graph allows designing codes with excellent performance in terms of error correction [14–16].

2.4 Decoding Algorithms for LDPC Codes

In this work we consider two types of MP decoding algorithms for LDPC codes.

Scaled Sum-Product Algorithm. We consider a BP decoding algorithm that generalizes of the classical sum-product algorithm (SPA), where the generalization introduces an attenuation of the extrinsic information produced at the CNs (see [17] for details). As we shall see, the attenuation is a heuristic to control the code performance at low error rates where trapping sets may lead to error floors.

Ternary Message Passing (TMP). TMP is an extension of binary message passing (BMP) decoding introduced in [18]. The exchanged messages between CNs and VNs belong to the ternary alphabet $\mathcal{M} = \{-1, 0, 1\}$, where 0 corresponds to an erasure. At the CNs the outgoing message is the product of the incoming messages. The update rule at the VNs involves weighting the channel and the incoming CN messages. The corresponding weights can be estimated from a DE analysis (see [19]). A quantization function is then applied to map the sum of the weighted messages to the ternary message alphabet $\mathcal{M}$.

2.5 Decoding of QC-LDPC codes with Hamming Weight Amplifiers

The decoding step in (4) using the parity-check matrix $\boldsymbol{H}$ is possible since $\boldsymbol{u}\boldsymbol{G}'\boldsymbol{Q}^T = \boldsymbol{x}'\boldsymbol{Q}^T$ is a codeword $\boldsymbol{x}$ of the LDPC code $\mathcal{C}$ described by $\boldsymbol{H}$ since

$$\boldsymbol{x}'\boldsymbol{H}'^T = \boldsymbol{x}'\boldsymbol{Q}^T\boldsymbol{H}^T = \boldsymbol{0} \quad \Longleftrightarrow \quad \boldsymbol{x}'\boldsymbol{Q}^T \in \mathcal{C}.$$

The error weight of transformed error $\boldsymbol{e}' = \boldsymbol{e}\boldsymbol{Q}^T$ in (4) is increased and upper bounded by

$$\mathsf{wht}\left(\boldsymbol{e}'\right) \leq e d_Q.$$

Due to the sparsity of $\boldsymbol{H}$, $\boldsymbol{Q}$ and $\boldsymbol{e}$ we have $\mathsf{wht}\left(\boldsymbol{e}'\right) \approx e d_Q$. In other words, the matrix $\boldsymbol{Q}$ increases the error weight and thus we call $\boldsymbol{Q}$ a *HWA*.

In the following we consider two decoding principles for LDPC codes with HWAs.

Basic Decoding Approach. A simple approach to decode an LDPC code with HWA is to decode the transformed ciphertext $\boldsymbol{u}\boldsymbol{G} = \mathrm{DEC}_{\boldsymbol{H}}(\tilde{\boldsymbol{c}})$ using a decoder for the LDPC code defined by $\boldsymbol{H}$ (see [6,7]). The sparse parity-check matrix $\mathrm{DEC}_{\boldsymbol{H}}(\cdot)$ has a good error-correction performance but the decoder must correct the amplified error $\boldsymbol{e}'$ of weight $\mathsf{wht}\left(\boldsymbol{e}'\right) \approx e d_Q$.

Decoding QC-LDPC-HWA Codes as QC-MDPC Codes. An alternative decoding approach is to consider $\boldsymbol{H}'$ as a parity-check matrix of a QC-MDPC code and decode the ciphertext $\boldsymbol{c}$ without using the transformation in (3):

$$\boldsymbol{u}\boldsymbol{G}' = \mathrm{DEC}_{\boldsymbol{H}'}(\boldsymbol{c}). \tag{5}$$

Compared to $\mathrm{DEC}_{\boldsymbol{H}}(\cdot)$, the error-correction performance of $\mathrm{DEC}_{\boldsymbol{H}'}(\cdot)$ is degraded due to the higher density of $\boldsymbol{H}'$. However, the decoder must only correct errors of weight $\mathsf{wht}(\boldsymbol{e}) = e$ (instead of ed_Q) since the Hamming weight is *not* increased by the transformation of the ciphertext $\boldsymbol{c}$ in (3).

Comparison of Decoding Strategies. In order to evaluate the performance of the previously described decoding strategies for LDPC codes with HWA, we analyze the error-correction capability using DE. The analysis, which addresses the performance of the relevant code ensembles in the asymptotic regime, i.e., as n goes to infinity, can be used to estimate the gains achievable in terms of error correction capability. We denote the iterative decoding threshold under SPA by $\delta^\star_{\mathsf{SPA}}$ and the decoding threshold of TMP by $\delta^\star_{\mathsf{TMP}}$. For a fair comparison, we consider the QC-MDPC ensemble for 80-bit security from [11] as a reference. For the reference ensemble, the estimate of the error-correction capability for a given n under SPA and TMP decoding can be roughly obtained as $n\delta^\star_{\mathsf{SPA}}$ and $n\delta^\star_{\mathsf{TMP}}$, respectively.

The parameters of the corresponding LDPC codes and the HWAs are chosen such that the row-weights of the resulting parity-check matrices $\boldsymbol{H}' = \boldsymbol{H}\boldsymbol{Q}$ match with the reference ensemble, i.e., we have $d_c' = d_c d_Q = 90$. In this setting, a rough estimate of the error-correction capability of the basic decoder is $n\delta^\star_{\mathsf{SPA}}/d_Q$ and $n\delta^\star_{\mathsf{TMP}}/d_Q$, where $\delta^\star_{\mathsf{SPA}}$ and $\delta^\star_{\mathsf{TMP}}$ are the decoding thresholds of the corresponding LDPC code under SPA and TMP decoding, respectively.

Table 1 shows, that decoding the MDPC code ($d_Q = 1$) gives a better error-correction performance than decoding the LDPC code from the amplified error.

Table 1. Comparison of decoding thresholds for basic (LDPC) and MDPC decoding.

Base matrix	n	d_Q	$n\delta^\star_{\mathsf{SPA}}/d_Q$	$n\delta^\star_{\mathsf{TMP}}/d_Q$
(45 45)	9602	1	113	113
(15 15)	9602	3	99	89
(9 9)	9602	5	87	78
(5 5)	9602	9	72	62

There is a bit-flipping-based [10] decoder, called "Q-decoder", that incorporates the knowledge of the HWA matrix $\boldsymbol{Q}$ during the decoding process [8]. The Q-decoder is equivalent to the bit-flipping decoder of the corresponding MDPC code [8, Lemma 1.5.1] but has a significantly lower computational complexity [8, Lemma 1.5.2].

3 Improved Protograph-Based Decoding of LDPC Codes with Hamming Weight Amplifiers

Motivated by the observations above, we derive a new protograph-based decoding framework for LDPC codes with HWAs that incorporates the knowledge about the HWA matrix $\boldsymbol{Q}$ at the receiver. The decoding framework allows to apply known MP decoding algorithms and has an improved error-correction capability compared to the naive approach and a significantly reduced computational complexity compared to the corresponding MDPC decoding approach (see (5)).

3.1 Protograph Representation of LDPC-HWA Decoding

Let $\mathcal{C}$ be an LDPC code with parity-check matrix $\boldsymbol{H}$ and let $\mathcal{C}'$ denote the code with parity-check matrix

$$\boldsymbol{H}' = \boldsymbol{H}\boldsymbol{Q}. \tag{6}$$

We have

$$\boldsymbol{x}\boldsymbol{H}^T = \boldsymbol{0}, \quad \forall \boldsymbol{x} \in \mathcal{C} \qquad \text{and} \qquad \boldsymbol{x}'\boldsymbol{H}'^T = \boldsymbol{0}, \quad \forall \boldsymbol{x}' \in \mathcal{C}'. \tag{7}$$

Using (6) we can rewrite (7) as

$$\boldsymbol{x}'\boldsymbol{H}'^T = \boldsymbol{x}'(\boldsymbol{H}\boldsymbol{Q})^T = \boldsymbol{x}'\boldsymbol{Q}^T\boldsymbol{H}^T = \boldsymbol{0}, \quad \forall \boldsymbol{x}' \in \mathcal{C}'.$$

Hence, $\boldsymbol{x}'\boldsymbol{Q}^T$ must be contained in $\mathcal{C}$ for all $\boldsymbol{x}' \in \mathcal{C}'$. Defining $\boldsymbol{x} = \boldsymbol{x}'\boldsymbol{Q}^T$ we can restate (7) as

$$\begin{aligned} \boldsymbol{x} &= \boldsymbol{x}'\boldsymbol{Q}^T \\ \boldsymbol{x}\boldsymbol{H}^T &= \boldsymbol{0} \end{aligned}$$

which we can write as $(\boldsymbol{x}' \ \boldsymbol{x})\,\boldsymbol{H}_{\text{ext}}^T = \boldsymbol{0}$ with

$$\boldsymbol{H}_{\text{ext}} = \begin{pmatrix} \boldsymbol{Q} & \boldsymbol{I}_{n\times n} \\ \boldsymbol{0} & \boldsymbol{H} \end{pmatrix}. \tag{8}$$

The matrix $\boldsymbol{H}_{\text{ext}}$ in (8) is a $(n+r) \times 2n$ parity-check matrix of an LDPC code of length $2n$ and dimension $n-r$. The corresponding base matrix of $\boldsymbol{H}_{\text{ext}}$ is

$$\boldsymbol{B}_{\text{ext}} = \begin{pmatrix} \boldsymbol{B}_Q & \boldsymbol{I} \\ \boldsymbol{0} & \boldsymbol{B}_H \end{pmatrix}. \tag{9}$$

The extended parity-check $\boldsymbol{H}_{\text{ext}}$ can be used for decoding where the n rightmost bits are associated to the punctured VNs and the n leftmost bits are associated to the ciphertext $\boldsymbol{c}'$. As mentioned in Sect. 2.3, introducing state VNs in a code graph can improve the error-correction capability of the code significantly [14–16].

The protograph corresponding to the base matrix $\boldsymbol{B}_{\text{ext}}$ is depicted in Fig. 1.

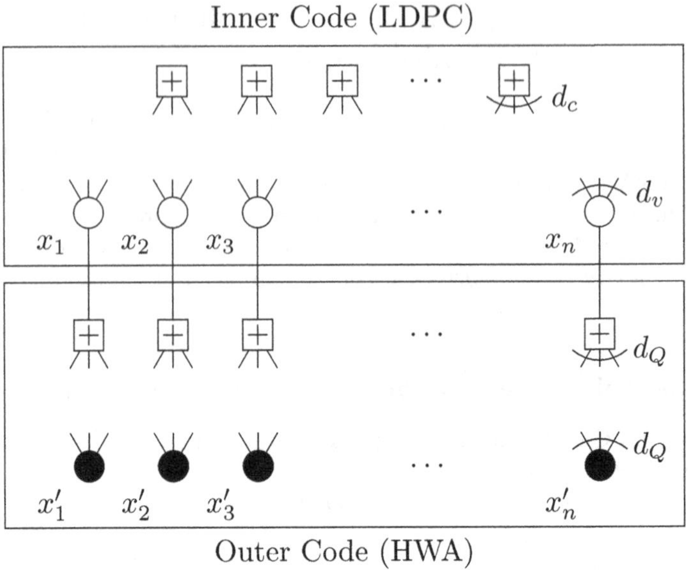

Fig. 1. Protograph representation of the LDPC code with HWA.

3.2 Complexity Considerations

The complexity of MP decoding depends on the CN and VN degrees of the underlying graph. Hence, the complexity of decoding the LDPC code is significantly lower than the complexity of decoding the corresponding MDPC code (see e.g. [7]). For the protograph-based decoding approach, we have $\mathcal{O}\left((d_c + d_Q)n\right)$ CN and VN messages per iteration whereas for MDPC decoding we have $\mathcal{O}\left(d_v d_Q n\right)$ CN and VN messages per iteration. Hence, the protograph-based decoding approach has a significantly lower complexity compared to the MDPC decoding approach.

Example 1. This effect was also observed in a Monte Carlo simulation for the first ensemble in Table 3. The simulation with a non-optimized ANSI C implementation of the TMP decoder took $36.9 \cdot 10^{-3}$ s/iteration for the protograph-based approach and 1.1 s/iteration for the corresponding MDPC approach.

4 Density Evolution Analysis

We now provide an asymptotic analysis of the code ensembles resulting from the different decoding approaches for LDPC codes with HWAs. The analysis is performed by means of DE under BP (SPA) and TMP decoding in order to get a rough estimate of the error correction capability of the codes drawn from the proposed ensembles.

4.1 BP: Quantized Density Evolution for Protographs

For BP, we resort to quantized DE (see [20,21] for details). The extension to protograph ensembles is straightforward and follows [16,22]. Simplified approaches based on Gaussian approximations are discarded due to the large CN degrees [23] of the MDPC ensembles.

4.2 TMP: Density Evolution for Protographs

The decoding threshold $\delta^\star_{\mathsf{TMP}}$, the optimal quantization threshold and the weights for the CN messages for TMP can be obtained by the DE analysis in [19].

4.3 Estimation of the Error Correction Capability

In order to evaluate the error correction performance of the above described decoding scheme we compare the protograph ensembles described by (9) with the corresponding QC-MDPC ensemble. As a reference we take the MDPC ensemble $\boldsymbol{B}_{\text{MDPC}} = (45\ 45)$ for 80-bit security from [11].

For a fair comparison, the reference ensembles in Table 2 are designed such that the base matrix $\boldsymbol{B}$ of $\boldsymbol{HQ}$ equals the base matrix $\boldsymbol{B}_{\text{MDPC}}$ of the corresponding QC-MDPC ensemble, i.e., we have

$$\boldsymbol{B}_H \boldsymbol{B}_Q \stackrel{!}{=} \boldsymbol{B}_{\text{MDPC}}.$$

For each ensemble, we computed the iterative decoding threshold, i.e., the largest channel error probability for which, in the limit of large n, DE predicts successful decoding convergence. We denote the iterative decoding threshold under SPA by $\delta^\star_{\mathsf{SPA}}$ and the decoding threshold of TMP by $\delta^\star_{\mathsf{TMP}}$. In Table 2 we provide a rough estimate of the number of errors at which the waterfall region of the block error probability is expected to be.

The result show, that for the considered parameters the error-correction capability of the protograph-based approach improves upon the MDPC-based approach under SPA decoding.

Table 2. Thresholds computed for different protographs.

Ensemble	Base matrix	n	$n\delta^\star_{\mathsf{SPA}}$	$n\delta^\star_{\mathsf{TMP}}$
$\mathscr{C}_{\mathsf{A}}$	$(45\ 45)$	9602	113	113
$\mathscr{C}_{\mathsf{B}}$	$\left(\begin{array}{cc\|cc} 2 & 1 & 1 & 0 \\ 1 & 2 & 0 & 1 \\ \hline 0 & 0 & 15 & 15 \end{array}\right)$	9602	121	103
$\mathscr{C}_{\mathsf{C}}$	$\left(\begin{array}{cc\|cc} 3 & 2 & 1 & 0 \\ 2 & 3 & 0 & 1 \\ \hline 0 & 0 & 9 & 9 \end{array}\right)$	9602	126	101
$\mathscr{C}_{\mathsf{D}}$	$\left(\begin{array}{cc\|cc} 5 & 4 & 1 & 0 \\ 4 & 5 & 0 & 1 \\ \hline 0 & 0 & 5 & 5 \end{array}\right)$	9602	127	80

4.4 Estimation of the Error Correction Capability for the LEDAcrypt Code Ensembles

Table 3 shows the decoding thresholds under SPA and TMP decoding for the protograph ensembles corresponding to the parameters in the current LEDAcrypt specifications [8, Table 3.1]. The decoding thresholds for the MDPC ensembles under SPA decoding could not be obtained by the quantized DE due to the high CN degrees. The results show that the proposed protograph-based approach has a similar error correction capability as the corresponding MDPC code. Further, the error correction capability under SPA and the efficient TMP decoding significantly improves upon the basic decoding approach [6, 7].

Table 3. Thresholds for different protographs for the parameters of LEDAcrypt [8, Table 3.1] for the NIST categories 1 (128 Bit), 3 (192 Bit) and 5 (256 Bit).

SL[Bit]	$\boldsymbol{B}_{ext}$	$\boldsymbol{B}_{MDPC}$	n	Proto		MDPC	Basic
				$n\delta^{\star}_{\mathsf{SPA}}$	$n\delta^{\star}_{\mathsf{TMP}}$	$n\delta^{\star}_{\mathsf{TMP}}$	$n\delta^{\star}_{\mathsf{SPA}}/d_Q$
128	$\left(\begin{array}{cc\|cc} 4 & 3 & 1 & 0 \\ 3 & 4 & 0 & 1 \\ \hline 0 & 0 & 11 & 11 \end{array}\right)$	(77 77)	29878	239	203	227	167
	$\left(\begin{array}{ccc\|ccc} 4 & 3 & 2 & 1 & 0 & 0 \\ 2 & 4 & 3 & 0 & 1 & 0 \\ 3 & 2 & 4 & 0 & 0 & 1 \\ \hline 0 & 0 & 0 & 9 & 9 & 9 \end{array}\right)$	(81 81 81)	23559	116	98	110	72
192	$\left(\begin{array}{cc\|cc} 5 & 3 & 1 & 0 \\ 3 & 5 & 0 & 1 \\ \hline 0 & 0 & 13 & 13 \end{array}\right)$	(104 104)	51386	316	272	303	222
	$\left(\begin{array}{ccc\|ccc} 4 & 4 & 3 & 1 & 0 & 0 \\ 3 & 4 & 4 & 0 & 1 & 0 \\ 4 & 3 & 4 & 0 & 0 & 1 \\ \hline 0 & 0 & 0 & 11 & 11 & 11 \end{array}\right)$	(121 121 121)	48201	162	144	163	106
256	$\left(\begin{array}{cc\|cc} 7 & 6 & 1 & 0 \\ 6 & 7 & 0 & 1 \\ \hline 0 & 0 & 11 & 11 \end{array}\right)$	(143 143)	73754	344	302	331	223
	$\left(\begin{array}{ccc\|ccc} 4 & 4 & 3 & 1 & 0 & 0 \\ 3 & 4 & 4 & 0 & 1 & 0 \\ 4 & 3 & 4 & 0 & 0 & 1 \\ \hline 0 & 0 & 0 & 15 & 15 & 15 \end{array}\right)$	(165 165 165)	82311	205	189	214	144

5 Simulation Results

In order to evaluate the error correction capability and validate the gains predicted by DE analysis, we performed Monte Carlo simulations for codes picked from the ensembles in Table 2. The number of iterations for the SPA and TMP algorithm was fixed to 100.

Figure 2 shows that the error correction performance of $\mathscr{C}_{\mathsf{B}}$ significantly improves upon the performance of $\mathscr{C}_{\mathsf{A}}$ under SPA decoding. For TMP decoding we observe, that we can recover much of the loss with respect to $\mathscr{C}_{\mathsf{A}}$. The figure also shows that the decoding thresholds predicted by DE (see Table 2) give a good estimate of the error correction performance gains.

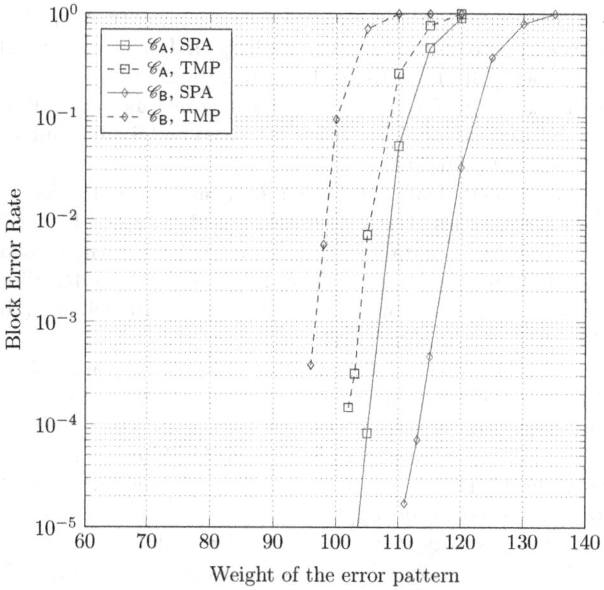

Fig. 2. Block error rate for codes from the ensembles in Table 2 under SPA and TMP decoding with 100 iterations.

6 Conclusions

In this paper, message passing (MP) decoding schemes for low density parity-check (LDPC) codes with Hamming weight amplifiers (HWAs), such as those used in the post-quantum crypto NIST proposal LEDAcrypt, were considered. A new protograph-based decoding framework that allows to analyze and optimize MP decoding schemes for LDPC codes with HWAs was presented. The new scheme uses a turbo-like principle to incorporate partial information about the errors that is available at the decoder and recovers most of the loss due to the error amplification of the HWA. Decoding thresholds for the resulting code ensembles under sum-product algorithm (SPA) and ternary message passing (TMP) decoding were obtained using density evolution (DE) analysis. The results show that the proposed decoding scheme improves upon the basic decoding approach and has a similar performance as the moderate-density parity-check (MDPC) decoding approach with a significantly lower complexity.

References

1. McEliece, R.J.: A public-key cryptosystem based on algebraic codes. Deep Space Netw. Progr. Rep. **44**, 114–116 (1978)
2. Rivest, R.L., Shamir, A., Adleman, L.: A method for obtaining digital signatures and public-key cryptosystems. Commun. ACM **21**(2), 120–126 (1978)

3. Monico, C., Rosenthal, J., Shokrollahi, A.: Using low density parity check codes in the McEliece cryptosystem. In: Proceedings IEEE International Symposium Information Theory (ISIT), Sorrento, Italy, p. 215 (2000)
4. Baldi, M., Chiaraluce, F.: Cryptanalysis of a new instance of McEliece cryptosystem based on QC-LDPC Codes. In: IEEE International Symposium on Information Theory, pp. 2591–2595 (2007)
5. Otmani, A., Tillich, J.-P., Dallot, L.: Cryptanalysis of two McEliece cryptosystems based on quasi-cyclic codes. Math. Comput. Sci. **3**(2), 129–140 (2010)
6. Baldi, M., Bodrato, M., Chiaraluce, F.: A new analysis of the mceliece cryptosystem based on QC-LDPC codes. In: Ostrovsky, R., De Prisco, R., Visconti, I. (eds.) SCN 2008. LNCS, vol. 5229, pp. 246–262. Springer, Heidelberg (2008). https://doi.org/10.1007/978-3-540-85855-3_17
7. Baldi, M., Bianchi, M., Chiaraluce, F.:Optimization of the parity-check matrix density in QC-LDPC code-based McEliece cryptosystems. In: 2013 IEEE International Conference on Communications Workshops (ICC), pp. 707–711. IEEE(2013)
8. Baldi, M., Barenghi, A., Chiaraluce, F., Pelosi, G., Santini, P.: LEDAcrypt: low-dEnsity parity-checkcoDe-bAsed cryptographic systems. NIST PQC submission (2019). https://www.ledacrypt.org/
9. Gallager, R.G.: Low-Density Parity-Check Codes. M.I.T. Press, Cambridge (1963)
10. Rudolph, L.: A class of majority logic decodable codes (corresp.). IEEE Trans. Inf. Theory **13**(2), 305–307 (1967)
11. Misoczki, R., Tillich, J.P., Sendrier, N., Barreto, P.S.L.M.: MDPC-McEliece: new McEliece variants from moderate density parity-check codes. In: IEEE International Symposium on Information Theory (ISIT), Istanbul, Turkey, pp. 2069–2073 (2013)
12. Kobara, K., Imai, H.: Semantically secure McEliece public-key cryptosystems - conversions for McEliece PKC. In: Kim, K. (ed.) PKC 2001. LNCS, vol. 1992, pp. 19–35. Springer, Heidelberg (2001). https://doi.org/10.1007/3-540-44586-2_2
13. Thorpe, J.: Low-density parity-check (LDPC) codes constructed from protographs. JPL IPN, Technical report, pp. 42–154, August 2003
14. Abbasfar, A., Yao, K., Disvalar, D.: Accumulate repeat accumulate codes. In: Proceedings of IEEE Globecomm, Dallas, Texas (2004)
15. Divsalar, D., Dolinar, S., Jones, C., Andrews, K.: Capacity-approaching protograph codes. IEEE JSAC **27**(6), 876–888 (2009)
16. Liva, G., Chiani, M.: Protograph LDPC code design based on EXIT analysis. In: Proceedings of IEEE Globecomm, Washington, US, pp. 3250–3254, December 2007
17. Bartz, H., Liva, G.: On decoding schemes for the MDPC-McEliece cryptosystem. In: 12th International ITG Conference on Systems, Communications and Coding (SCC), pp. 1–6. VDE (2019)
18. Lechner, G., Pedersen, T., Kramer, G.: Analysis and design of binary message passing decoders. IEEE Trans. Commun. **60**(3), 601–607 (2011)
19. Yacoub, E.B., Steiner, F., Matuz, B., Liva, G.: Protograph-based LDPC code design for ternary message passing decoding. In: 12th International ITG Conference on Systems, Communications and Coding, SCC 2019, pp. 1–6. VDE, February 2019
20. Chung, S.-Y., Forney, G.D., Richardson, T.J., Urbanke, R.: On the design of low-density parity-check codes within 0.0045 dB of the Shannon limit. IEEE Commun. Lett. **5**(2), 58–60 (2001)

21. Jin, H., Richardson, T.: A new fast density evolution. In: 2006 IEEE Information Theory Workshop-ITW Punta del Este, pp. 183–187. IEEE (2006)
22. Pulini, P., Liva, G., Chiani, M.: Unequal diversity LDPC codes for relay channels. IEEE Trans. Wirel. Commun. **12**(11), 5646–5655 (2013)
23. Chung, S.-Y., Richardson, T.J., Urbanke, R.L.: Analysis of sum-product decoding of low-density parity-check codes using a Gaussian approximation. IEEE Trans. Inf. Theory **47**(2), 657–670 (2001)

MURAVE: A New Rank Code-Based Signature with MUltiple RAnk VErification

Terry Shue Chien Lau(✉) and Chik How Tan

Temasek Laboratories, National University of Singapore,
5A Engineering Drive 1, #09-02, Singapore 117411, Singapore
{tsltlsc,tsltch}@nus.edu.sg

Abstract. We propose a new rank metric code-based signature scheme constructed via the Schnorr approach. Our scheme is designed in a way to avoid leakage of the information on the support for the secret key used in the signature generation. We define some new problems in rank metric code-based cryptography: the Rank Support Basis Decomposition problem and the Advanced Rank Support Basis Decomposition problem. We also discuss their hardness and solving complexity. Furthermore, we give a proof in the EUF-CMA security model, by reducing the security of our scheme to the Rank Syndrome Decoding problem, the Ideal LRPC Codes Indistinguishability problem and the Decisional Rank Support Basis Decomposition problem. We analyze the practical security for our scheme against the known attacks on rank metric signature schemes. Our scheme is efficient in terms of key size (5.33 KB) and of signature sizes (9.69 KB) at 128-bit classical security level.

Keywords: Rank metric · Digital signature scheme · Provable security · Code-based cryptography

1 Introduction

In 2019, the National Institute of Standards and Technology (NIST) has announced second round candidates in the post-quantum standardization process. None of the code-based digital signature schemes (pqsigRM [25], RaCoSS [29] and RankSign [6]) were selected as the signature candidates in second round. In particular, both RaCoSS and RankSign were cryptanalyzed in [13,16] respectively, while pqsigRM has large key size comparing to other signature scheme submissions. Recently, Debris-Alazard et al. [15] has proposed a code-based signature scheme in Hamming metric, namely Wave, constructed via a code-based one way trapdoor function that meets the preimage sampleable property. Although it is still secure up-to-date, it requires public key size of 3 MB with signature size of around 1.6 KB for 128-bit security. Constructing a secure signature scheme with compact key sizes still remains as a challenging problem in code-based cryptography.

M. Baldi et al. (Eds.): CBCrypto 2020, LNCS 12087, pp. 94–116, 2020.
https://doi.org/10.1007/978-3-030-54074-6_6

Besides code-based signature constructed via the Hash-and-Sign approach (for instances RankSign and Wave), there are two other approaches in constructing code-based signature in the rank metric settings. The first approach considers a zero-knowledge authentication algorithm and applies Fiat-Shamir transformation to convert it into a signature scheme. Signature schemes such as Rank CVE and Veron [9], cRVDC [10] were constructed via this approach. Although such signatures are efficient in terms of public key size and secret key size (about one thousand bits), they suffered a major drawback in terms of signature size (up to two hundred thousand bits) as the number of necessary rounds for the protocol is large.

On the other hand, the second approach adapts the Schnorr signature scheme [30] into rank metric context. Signature schemes such as TPL [32] and RQCS [31] were proposed via this approach. The signature of these schemes consists of a vector $\boldsymbol{z} = \boldsymbol{y} + \boldsymbol{c} \circ \boldsymbol{x}$, where $\boldsymbol{y}$ is an ephemeral key, $\boldsymbol{c}$ is the output of a hash function commited with a message to be signed, $\boldsymbol{x}$ is the secret key of the signature scheme and $\circ$ is an operation on $\boldsymbol{c}$ and $\boldsymbol{x}$. In the verification step, one needs to verify that $\mathrm{rk}(\boldsymbol{z}) \leq \mathrm{rk}(\boldsymbol{y}) + \mathrm{rk}(\boldsymbol{c}) \times \mathrm{rk}(\boldsymbol{x})$. Although these schemes constructed via the Schnorr approach have compact key size and signature size, the main drawback of the signature is the leakage of the information in the signature to recover the secret key. Lau et al. [24] generalized the attack on RQCS in [2] and proposed a generic key recovery attack (denoted as LTP attack), which can be applied to cryptanalyze TPL signature schemes within a few seconds. In particular, an adversary can exploit the information from $\boldsymbol{z}$ and $\boldsymbol{c}$ to recover a support basis for the secret key $\boldsymbol{x}$. With the recovered basis, the adversary can compute a support matrix for the secret key using the information on public key.

In 2019, Aragon et al. [3] proposed a signature scheme called Durandal which overcomes the drawback of rank metric signatures constructed via the Schnorr approach. Let W and V be subspaces of $\mathbb{F}_{q^m}$. Instead of having a signature $\boldsymbol{z}$ in the form of $\boldsymbol{z} = \boldsymbol{y} + \boldsymbol{c} \circ \boldsymbol{x} \in (W + V)^n$, Durandal includes an extra secret $\boldsymbol{x}'$ in the signature so that $\boldsymbol{z} = \boldsymbol{y} + \boldsymbol{c} \circ \boldsymbol{x} + \boldsymbol{p} \circ \boldsymbol{x}' \in (W + U)^n$, where U is a filtered subspace of V. Such U is filtered in a way that an adversary is not able to exploit the information from $\boldsymbol{z}$, $\boldsymbol{c}$ and $\boldsymbol{p}$ to recover a support basis for the secret key $\boldsymbol{x}$ and $\boldsymbol{x}'$, hence LTP attack is difficult to be applied on the Durandal signature. Although Durandal has good performance in the compactness of public key size (of 15.25 KB) and signature size (4.06 KB), it has the following challenges in implementation:

1. Durandal is a signature scheme that applies the "Fiat-Shamir with Aborts" (FS Aborts) strategy [27]. In particular, in the offline signature generation phase, the probability that a random U satisfies the required filtered conditions is approximately $e^{-2} \approx 0.135$. As a consequence, the process of randomizing U has to be repeated 7.5 times on average to obtain a suitable subspace U.
2. In the offline signature generation phase, it is required to compute an $m_d \times m_d$-matrix D, which would be required for the computation of the signature in the online phase. The computation of D requires inverting a linear system

with m_d equations, hence the cost is $O\left(m_d^\omega\right)$ multiplications in the underlying base field, where ω is the constant of linear algebra.

As a consequence, this results in longer signature generation timing, especially during the offline phase.

Our Contribution. In this paper, we propose a new signature scheme constructed via the Schnorr approach. We propose a new method to avoid leakage of the information on the support for the secret key used in the signature generation, hence able to resist against LTP attack.

In particular, let $\mathcal{U}$, $\mathcal{V}$ and $\mathcal{F}$ be random subspaces of $\mathbb{F}_{q^m}$ with dimensions r_u, r_v and r_f respectively. We consider $\mathcal{U}$, $\mathcal{V}$ and $\mathcal{F}$ to be the secret key of our signature scheme. Let $\boldsymbol{b}$ be a random q^m-ary vector of rank r_b, $\boldsymbol{\lambda}$ be a random q^m-ary vector of rank r_λ, and $\boldsymbol{f}$ be a random vector in $\mathcal{F}$. Then, compute $\boldsymbol{z} = \boldsymbol{b} + \boldsymbol{\lambda}(\boldsymbol{v}'' - \boldsymbol{u}''\boldsymbol{f})$ where $\boldsymbol{u}''$ and $\boldsymbol{v}''$ are random vectors in $\mathcal{U}$ and $\mathcal{V}$ respectively. Although $\boldsymbol{z}$ is sent out as signature, the support bases of the secret $\mathcal{U}$, $\mathcal{V}$ and $\mathcal{F}$ are masked with $\boldsymbol{\lambda}$ which is not made public. Our signature scheme has the following advantages:

1. We use only random vectors $\boldsymbol{\lambda}$ to mask the secret key. Our masking only involves polynomial additions and multiplications and does not require any costly computation (such as solving a linear system).
2. Each steps in our signature generation is deterministic, i.e., we are not required to apply the Fiat-Shamir with Aborts strategy. As such, our signature generation can be much more efficient.
3. Our signature scheme has the most compact public key size (of 5.33 KB) among all the rank metric signature schemes constructed via the Schnorr approach.

In addition, we define some new problems in rank metric code-based cryptography: the Rank Support Basis Decomposition problem and the Advanced Rank Support Basis Decomposition problem. We also discuss their hardness and solving complexity. Furthermore, we give a proof in the EUF-CMA security model, by reducing the security of our scheme to the Rank Syndrome Decoding problem, the Ideal LRPC Codes Indistinguishability problem and the Advanced Rank Support Basis Decomposition problem. We also analyze the practical security for our scheme against the known attacks on rank metric signature schemes.

Organization of the Paper. This paper is organized as follows: Sect. 2 reviews some definitions and preliminary results in rank metric coding theory. Section 3 presents the specification for our new signature scheme, MURAVE. In Sect. 4, we define some new problems in rank metric code-based cryptography with discussions on their hardness and solving complexity. We prove that MURAVE achieves EUF-CMA security in Sect. 5. We also show that MURAVE is secure against the existing attacks on rank metric signature schemes. Section 6 presents the parameters and the implementations timing for MURAVE. We conclude our paper in Sect. 7.

2 Rank Metric Codes

In this section, we introduce some basic definitions and preliminaries in rank metric coding theory. Let q be a prime power. We denote $\mathbb{F}_q$ as the finite field with q elements, and $\mathbb{F}_{q^m}$ as the finite field with q^m elements. We can view $\mathbb{F}_{q^m}$ as an m-dimensional vector space over $\mathbb{F}_q$ with basis $\{\beta_1, \ldots, \beta_m\}$, i.e., $\mathbb{F}_{q^m} = \langle \beta_1, \ldots, \beta_m \rangle_{\mathbb{F}_q}$ where $\langle \beta_1, \ldots, \beta_m \rangle_{\mathbb{F}_q}$ is the $\mathbb{F}_q$-linear span of the elements in $\{\beta_1, \ldots, \beta_m\}$. We will omit the term $\mathbb{F}_q$ if there is no ambiguity on $\mathbb{F}_q$.

Definition 1. An $[n, k]$-linear code $\mathcal{C}$ of length n and dimension k is a linear subspace of $\mathbb{F}_{q^m}^n$ with dimension k. Moreover, $\mathcal{C}$ can be represented by a generator matrix $G \in \mathbb{F}_{q^m}^{k \times n}$ with $\text{rk}(G) = k$, or by a parity-check matrix $H \in \mathbb{F}_{q^m}^{(n-k) \times n}$ with $\text{rk}(H) = n - k$ and $GH^T = \mathbf{0}$. More specifically,

$$\mathcal{C} = \left\{ \boldsymbol{x} \in \mathbb{F}_{q^m}^n : \boldsymbol{x} = \boldsymbol{m}G, \text{ for all } \boldsymbol{m} \in \mathbb{F}_{q^m}^k \right\} = \left\{ \boldsymbol{x} \in \mathbb{F}_{q^m}^n : \boldsymbol{x}H^T = \mathbf{0} \right\}.$$

We say that G (respectively H) is in systematic form if it is of the form $[I_k \mid A]$ where $A \in \mathbb{F}_{q^m}^{k \times (n-k)}$ (respectively $[I_{n-k} \mid B]$ where $B \in \mathbb{F}_{q^m}^{(n-k) \times k}$).

Definition 2. Let $\boldsymbol{x} = (x_1, \ldots, x_n) \in \mathbb{F}_{q^m}^n$ and $\boldsymbol{\beta} = (\beta_1, \ldots, \beta_m) \in \mathbb{F}_{q^m}^m$ where $\{\beta_1, \ldots, \beta_m\}$ is a basis of $\mathbb{F}_{q^m}$. For $1 \leq i \leq n$, we can write $x_i = \sum_{j=1}^{m} c_{ji} \beta_j$ where $c_{ji} \in \mathbb{F}_q$, i.e., $\boldsymbol{x} = \boldsymbol{\beta} C$ where $C = [c_{ji}]_{\substack{1 \leq j \leq m, \\ 1 \leq i \leq n}} \in \mathbb{F}_q^{m \times n}$. The rank weight $\text{rk}(\cdot)$ of $\boldsymbol{x}$ is defined as $\text{rk}(\boldsymbol{x}) := \text{rk}(C)$.

The following is a well known result for a vector $\boldsymbol{x}$ of rank r:

Lemma 1 ([23, Proposition 10]). Let $\boldsymbol{x} = (x_1, \ldots, x_n) \in \mathbb{F}_{q^m}^n$ be a vector such that $\text{rk}(\boldsymbol{x}) = r$. Then there exist $\hat{\boldsymbol{x}} = (\hat{x}_1, \ldots, \hat{x}_r) \in \mathbb{F}_{q^m}^r$ and $E_x \in \mathbb{F}_q^{r \times n}$ such that $\boldsymbol{x} = \hat{\boldsymbol{x}} E_x$ with $\text{rk}(\hat{\boldsymbol{x}}) = r$ and $\text{rk}(E_x) = r$. We call $\text{supp}(\boldsymbol{x}) = \langle x_1, \ldots, x_n \rangle \subset \mathbb{F}_{q^m}^n$ as the support for $\boldsymbol{x}$, E_x as a support matrix for $\boldsymbol{x}$, and $\{\hat{x}_1, \ldots, \hat{x}_r\}$ as a support basis for $\boldsymbol{x}$.

Notation. The following are the notations used throughout this paper:

- By abuse of notation, we can view a vector $\boldsymbol{g} = (g_0, \ldots, g_{k-1}) \in \mathbb{F}_{q^m}^k$ as a polynomial $G(X) = \sum_{i=0}^{k-1} g_i X^i$.
- Denote $E_{m,n,r} := \left\{ \boldsymbol{x} : \boldsymbol{x} \in \mathbb{F}_{q^m}^n, \text{rk}(\boldsymbol{x}) = r \right\}$.
- Let $\boldsymbol{g}_1, \boldsymbol{g}_2 \in \mathbb{F}_{q^m}^k$ and $P(x) \in \mathbb{F}_q[X]$ be an irreducible polynomial of degree k. Let $G_1(X)$ and $G_2(X)$ be the polynomials associated respectively to $\boldsymbol{g}_1$ and $\boldsymbol{g}_2$. We denote $\boldsymbol{g}_1 \boldsymbol{g}_2 \mod P := G_1(X) G_2(X) \mod P$. We will often omit mentioning the polynomial P if there is no ambiguity.
- Denote $\mathbf{1} := (1, 0, ..., 0) \in \mathbb{F}_{q^m}^k$.
- Denote $\boldsymbol{g}^{-1} \in \mathbb{F}_{q^m}^k$ as the vector (or polynomial) such that $\mathbf{1} = \boldsymbol{g}\boldsymbol{g}^{-1} \mod P$. If $\boldsymbol{g}^{-1}$ exists, then $\boldsymbol{g}$ is invertible.

- Let $V = \text{supp}(\boldsymbol{v})$. Denote $V^{-1} := \text{supp}(\boldsymbol{v}^{-1})$.
- Let X be a finite set. We write $x \xleftarrow{\$} X$ to denote the assignment to x of an element randomly sampled from the uniform distribution on X.
- Denote $\mathsf{SubSp}(d, \mathbb{F}_{q^m})$ as the set of all d-dimensional $\mathbb{F}_q$-subspaces of $\mathbb{F}_{q^m}$.
- Let $A = \langle a_1, \ldots, a_r \rangle \in \mathsf{SubSp}(r, \mathbb{F}_{q^m})$ and $B = \langle b_1, \ldots, b_d \rangle \in \mathsf{SubSp}(d, \mathbb{F}_{q^m})$. Denote the product space $A.B := \langle a_1 b_1, \ldots, a_r b_d \rangle \subset \mathbb{F}_{q^m}$ as a vector subspace with dimension at most rd.
- Denote $\mathsf{RV}(r, \mathbb{F}_{q^m}, k)$ as an algorithm that outputs a random r-dimensional vector subspace $V \xleftarrow{\$} \mathsf{SubSp}(r, \mathbb{F}_{q^m})$ and a random k-dimensional vector $\boldsymbol{v} \xleftarrow{\$} V^k$.

We consider $[2k, k]$ ideal codes defined as follows:

Definition 3 (Ideal Codes). Let $P(X) \in \mathbb{F}_q[X]$ be an irreducible polynomial of degree k and $\boldsymbol{g_1}, \boldsymbol{g_2} \in \mathbb{F}_{q^m}^k$. For $1 \leq j \leq 2$, let $G_j(X) = \sum_{i=0}^{k-1} g_{ji} X^i$ be the polynomials associated respectively to $\boldsymbol{g_j} = (g_{j,0}, \ldots, g_{j,k-1})$. The $[2k, k]$ ideal code $\mathcal{C}$ with generator $(\boldsymbol{g_1}, \boldsymbol{g_2})$ is the code with generator matrix

$$G = \left[\begin{array}{c|c} X^0 G_1(X) \bmod P & X^0 G_2(X) \bmod P \\ \vdots & \vdots \\ X^{k-1} G_1(X) \bmod P & X^{k-1} G_2(X) \bmod P \end{array}\right]. \tag{1}$$

Equivalently, we have $\mathcal{C} = \left\{(\boldsymbol{x}\boldsymbol{g_1} \bmod P, \boldsymbol{x}\boldsymbol{g_2} \bmod P) : \text{ for all } \boldsymbol{x} \in \mathbb{F}_{q^m}^k\right\}$. Furthermore, if $\boldsymbol{g_1}$ is invertible, we may express the code in systematic form, i.e. $\mathcal{C} = \left\{(\boldsymbol{x}, \boldsymbol{x}\boldsymbol{g}) : \text{ for all } \boldsymbol{x} \in \mathbb{F}_{q^m}^k\right\}$ where $\boldsymbol{g} = \boldsymbol{g_1}^{-1}\boldsymbol{g_2} \bmod P$. We call $\boldsymbol{g}$ and P as the generator for this code $\mathcal{C}$.

Remark 1. Let $\mathcal{C}$ be a $[2k, k]$ ideal code with generator $(\boldsymbol{g_1}, \boldsymbol{g_2})$. We say that $(\boldsymbol{h}_1, \boldsymbol{h}_2)$ and P define a parity check matrix of $\mathcal{C}$ if $H = [H_1 \mid H_2]$ is a parity check matrix of G as defined in (1) where

$$H_1 = \begin{bmatrix} X^0 \boldsymbol{h}_1 \bmod P \\ \vdots \\ X^{k-1} \boldsymbol{h}_1 \bmod P \end{bmatrix} \quad \text{and} \quad H_2 = \begin{bmatrix} X^0 \boldsymbol{h}_2 \bmod P \\ \vdots \\ X^{k-1} \boldsymbol{h}_2 \bmod P \end{bmatrix}.$$

Similarly, if $\boldsymbol{h}_1^{-1}$ is invertible, we call $\boldsymbol{h} = \boldsymbol{h}_1^{-1}\boldsymbol{h}_2$ and P as the generator for the parity check matrix of this ideal code $\mathcal{C}$.

The following is a problem defined in rank metric coding theory.

Problem 1 (Rank Syndrome Decoding (RSD) Problem). Let H be a full rank $(n-k) \times n$ matrix over $\mathbb{F}_{q^m}$, $\boldsymbol{s} \in \mathbb{F}_{q^m}^{n-k}$ and r an integer. The Rank Syndrome Decoding problem $\mathsf{RSD}_H(q, m, n, k, r)$ is to determine a vector $\boldsymbol{x} \in E_{m,n,r}$ such that $\boldsymbol{x}H^T = \boldsymbol{s}$.

The RSD problem is analogous to the classical syndrome decoding SD problem in Hamming metric (SD was shown to be an NP-complete problem [11]). Recently, Gaborit and Zémor [19] showed that if there were efficient probabilistic algorithms for solving the RSD problem, then there would exist efficient probabilistic algorithms to solve the syndrome decoding problem in Hamming metric. As a result, the RSD problem is accepted by the research community as a hard problem for which a rank metric code-based cryptosystem is based on.

There are generally two types of generic attacks on the RSD problem: combinatorial attack and algebraic attack.

Combinatorial Attack. The combinatorial approach depends on counting the number of possible support basis of size r or support matrix of rank r for a rank code of length n over $\mathbb{F}_{q^m}$, which corresponds to the number of subspaces of dimension r in $\mathbb{F}_{q^m}$. We summarize the existing combinatorial attacks with their complexities in Table 1.

Table 1. Combinatorial attacks on RSD with their corresponding solving complexities

Attacks	Complexity
CS [14]	$O\left((nr+m)^3 q^{(m-r)(r-1)}\right)$
GRS-I [18]	$\begin{cases} O\left((n-k)^3 m^3 q^{r\min\{k,\lfloor\frac{km}{n}\rfloor\}}\right) & \text{if } s \neq \mathbf{0}, \\ O\left((n-k)^3 m^3 q^{(r-1)\min\{k,\lfloor\frac{km}{n}\rfloor\}}\right) & \text{if } s = \mathbf{0}. \end{cases}$
OJ-I [28]	$O\left(r^3 m^3 q^{(r-1)(k+1)}\right)$
OJ-II [28]	$O\left((k+r)^3 r^3 q^{(m-r)(r-1)}\right)$
GRS-II [18]	$O\left((n-k)^3 m^3 q^{(r-1)\min\{k+1,\frac{(k+1)m}{n}\}}\right)$
AGHT [5]	$O\left((n-k)^3 m^3 q^{r\frac{(k+1)m}{n}-m}\right)$

Algebraic Attack. The nature of the rank metric favors algebraic attacks using Gröbner bases, as they are largely independent of the value q. There are mainly four approaches in translating the notion of rank into algebraic setting. The first approach considers directly the RSD problem and was introduced by Levy-dit-Vehel and Perret [26]. In particular, they use Gröbner basis techniques to solve the polynomial system arising in the Ourivski-Johansson algebraic modeling [28]. However, the complexity of solving the quadratic system from their attack is hard to evaluate, especially when $r \geq 4$. Recently, Bardet et al. [7] followed this approach and show that this polynomial system can be augmented with additional equations that are easy to compute and bring on a substantial speed-up in the Gröbner basis computation for solving the system. The second approach reduces RSD problem into MinRank problem [17], but such reduction only works for certain type of MinRank parameters and not for usual parameters used with rank codes based cryptography. The third approach is proposed by Gaborit et al. [18] by considering the linearized q-polynomials in solving the RSD problem. More recently, Bardet et al. [8] followed the approach in [7] and proposed a new modeling to solve the RSD problem. This new modeling avoids the use of Gröbner basis algorithms and brings on a substantial speed-up in the computations.

We summarize the existing algebraic attacks and their complexities in Table 2, with the following notations:

- the constant of linear algebra ω, with value $\omega \approx 2.807$
- an integer $a \geq 0$
- $\mathsf{Bin}_r^n := \binom{n}{r}$
- $p := \max\left\{i : m\mathsf{Bin}_r^{n-i-k-1} \geq \mathsf{Bin}_r^{n-i} - 1\right\}$
- $A_t := \sum_{j=1}^{t} \mathsf{Bin}_r^n \mathsf{Bin}_j^{mk+1}$
- $B_t := \sum_{j=1}^{t} \left(m\mathsf{Bin}_r^{n-k-1}\mathsf{Bin}_j^{mk+1} + \sum_{i=1}^{j} (-1)^{i+1}\mathsf{Bin}_{r+i}^n \mathsf{Bin}_i^{m+i-1}\mathsf{Bin}_{j-i}^{mk+1} \right)$.
- $b := \min\{t : 0 < t < r+2, A_t \leq B_t\}$

Table 2. Conditions and complexities of algebraic attacks on $\mathsf{RSD}(q, m, n, k, r)$.

Attacks	Conditions	Complexity
FLP [17]	$m = n,\ (n-r)^2 = nk$	$O\left((\log q) n^{3(n-r)^2}\right)$
CGK [21]		$O\left(k^3 m^3 q^{r\left\lceil \frac{km}{n} \right\rceil}\right)$
GRS [18]	$(r+1)(k+1) - (n+1) \leq 0$	$O\left(((r+1)(k+1)-1)^3\right)$
	$\left\lceil \frac{(r+1)(k+1)-(n+1)}{r} \right\rceil \leq k$	$O\left(r^3 k^3 q^{r\left\lceil \frac{(r+1)(k+1)-(n+1)}{r} \right\rceil}\right)$
BBB+ [7]	$m\mathsf{Bin}_r^{n-k-1} \geq \mathsf{Bin}_r^n$	$O\left(\left(\frac{((m+n)r)^r}{r!}\right)^\omega\right)$
	$m\mathsf{Bin}_r^{n-k-1} < \mathsf{Bin}_r^n$	$O\left(\left(\frac{((m+n)r)^{r+1}}{(r+1)!}\right)^\omega\right)$
BBC+ [8]	$m\mathsf{Bin}_r^{n-p-k-1} \geq \mathsf{Bin}_r^{n-p} - 1$	$O\left(m\mathsf{Bin}_r^{n-p-k-1}\mathsf{Bin}_r^{n-p\omega-1}\right)$
	$m\mathsf{Bin}_r^{n-k-1} \geq \mathsf{Bin}_r^{n-a} - 1$	$O\left(q^{ar} m\mathsf{Bin}_r^{n-k-1}\mathsf{Bin}_r^{n-a\omega-1}\right)$
	$A_b - 1 \leq B_b,\ q = 2$	$O\left(B_b A_b^{\omega-1}\right)$

3 MURAVE: A New Signature Scheme

The following is the specification for our new signature scheme. We name our signature scheme as MURAVE, as we need to perform multiple rank verification for different components of the signature in the verification phase.

Setup: Generate global parameters with integers q, m, l, k, r_λ, r_u, r_v, r_e, r_f, r_w where q is a prime power. Output parameters $\mathsf{param} = (q, m, l, k, r_\lambda, r_u, r_v, r_e, r_f, r_w)$.

Key.Gen(param): Let $\mathcal{H}_{\mathsf{M},\mathsf{N}} : \mathsf{M} \to \mathsf{N}$ be a collision-resistant hash function where

$$\mathsf{M} = \left(\mathbb{F}_{q^m}^{3k}\right)^l \times \{0,1\}^* \times \mathbb{F}_{q^m}^k \times \left(\mathbb{F}_{q^m}^{2k}\right)^l, \quad \mathsf{N} = \mathbb{F}_q^{kl}.$$

Choose random $\boldsymbol{h} \overset{\$}{\leftarrow} \mathbb{F}_{q^m}^k$. For $1 \le i \le l$:

1. $\boldsymbol{\eta}_i \overset{\$}{\leftarrow} \mathbb{F}_q^k$,
2. $(\boldsymbol{\mu}_i, \mathcal{U}_i) \overset{\$}{\leftarrow} \mathsf{RV}(r_u, \mathbb{F}_{q^m}, k)$ and $(\boldsymbol{v}_i, \mathcal{V}_i) \overset{\$}{\leftarrow} \mathsf{RV}(r_v, \mathbb{F}_{q^m}, k)$,
3. $(\boldsymbol{e}_i, \mathcal{E}_i) \overset{\$}{\leftarrow} \mathsf{RV}(r_e, \mathbb{F}_{q^m}, k)$ and $(\boldsymbol{w}_i, \mathcal{W}_i) \overset{\$}{\leftarrow} \mathsf{RV}(r_w, \mathbb{F}_{q^m}, k)$ such that $\boldsymbol{e}_i$, $\boldsymbol{w}_i$ are invertible,
4. $(\boldsymbol{f}_i, \mathcal{F}_i) \overset{\$}{\leftarrow} \mathsf{RV}(r_f, \mathbb{F}_{q^m}, k)$ such that $1 \notin \mathcal{F}_i$,.

Compute $\boldsymbol{s}_i = \boldsymbol{w}_i(\boldsymbol{\eta}_i \boldsymbol{\mu}_i + \boldsymbol{v}_i \boldsymbol{h})$ and $\boldsymbol{s}'_i = \boldsymbol{e}_i + \boldsymbol{e}_i \boldsymbol{f}_i \boldsymbol{h}$. Output $(\mathsf{pk}, \mathsf{sk})$ where

$$\mathsf{pk} = \left(\boldsymbol{h}, \{\boldsymbol{s}_i, \boldsymbol{s}'_i\}_{1 \le i \le l}\right), \quad \mathsf{sk} = \left(\{\boldsymbol{w}_i, \boldsymbol{\mu}_i, \boldsymbol{\eta}_i, \boldsymbol{v}_i, \boldsymbol{e}_i, \boldsymbol{f}_i\}_{1 \le i \le l}\right).$$

Sign($\boldsymbol{m}$, pk, sk): Let $\boldsymbol{m}$ be a message to be signed. For $1 \le i \le l$:

- $(\boldsymbol{a}_i, A_i) \overset{\$}{\leftarrow} \mathsf{RV}(r_a, \mathbb{F}_{q^m}, k)$ and $(\boldsymbol{b}_i, B_i) \overset{\$}{\leftarrow} \mathsf{RV}(r_b, \mathbb{F}_{q^m}, k)$,
- $\boldsymbol{\eta}'_i \overset{\$}{\leftarrow} \mathbb{F}_q^k$ and $\boldsymbol{v}'_i \overset{\$}{\leftarrow} \mathcal{V}^k$,
- $(\boldsymbol{\lambda}_i, \Lambda_i) \overset{\$}{\leftarrow} \mathsf{RV}(r_\lambda, \mathbb{F}_{q^m}, k)$ such that $\mathbf{1} \notin \Lambda_i$.

Compute

1. $\boldsymbol{\tau}_i = \boldsymbol{\lambda}_i \boldsymbol{w}_i^{-1}$, $\boldsymbol{d}_i = -\boldsymbol{\lambda}_i \boldsymbol{\mu}_i \boldsymbol{e}_i^{-1}$, $\gamma_i = \boldsymbol{a}_i + \boldsymbol{\lambda}_i \boldsymbol{\eta}'_i \boldsymbol{\mu}_i + (\boldsymbol{b}_i + \boldsymbol{\lambda}_i \boldsymbol{v}'_i)\boldsymbol{h}$,
2. $\boldsymbol{c} = (\boldsymbol{c}_1, \ldots, \boldsymbol{c}_l) = \mathcal{H}\left(\{\gamma_i, \boldsymbol{\tau}_i, \boldsymbol{d}_i\}_{1 \le i \le l}, \boldsymbol{m}, \mathsf{pk}\right)$ where $\boldsymbol{c}_i \in \mathbb{F}_q^k$.
3. $\boldsymbol{\rho}_i = \boldsymbol{\eta}'_i + \boldsymbol{c}_i \boldsymbol{\eta}_i$, $\boldsymbol{u}''_i = \boldsymbol{\rho}_i \boldsymbol{\mu}_i$, $\boldsymbol{v}''_i = \boldsymbol{v}'_i + \boldsymbol{c}_i \boldsymbol{v}_i$, $\boldsymbol{z}_i = \boldsymbol{b}_i + \boldsymbol{\lambda}_i(\boldsymbol{v}''_i - \boldsymbol{u}''_i \boldsymbol{f}_i)$.

Output signature $\boldsymbol{\sigma} = \left(\{\boldsymbol{c}_i, \boldsymbol{a}_i, \boldsymbol{z}_i, \boldsymbol{\tau}_i, \boldsymbol{d}_i, \boldsymbol{\rho}_i\}_{1 \le i \le l}\right)$.

Verify($\boldsymbol{\sigma}$, pk): Accept a signature $\boldsymbol{\sigma}$ if and only if

1. For each $1 \le i \le l$, $\mathrm{rk}(\boldsymbol{\tau}_i) = \mathrm{rk}(\boldsymbol{d}_i) = k$, $\mathrm{rk}(\boldsymbol{a}_i) = r_a$ and $\mathrm{rk}(\boldsymbol{z}_i) \le r_b + r_\lambda(r_v + r_u r_f)$.
2. For each $1 \le i \le l$, $\boldsymbol{\rho}_i \in \mathbb{F}_q^k$.
3. $\boldsymbol{c} = \mathcal{H}\left(\{\gamma_i, \boldsymbol{\tau}_i, \boldsymbol{d}_i\}_{1 \le i \le l}, \boldsymbol{m}, \mathsf{pk}\right)$ where $\gamma_i = \boldsymbol{a}_i + \boldsymbol{z}_i \boldsymbol{h} - \boldsymbol{c}_i \boldsymbol{\tau}_i \boldsymbol{s}_i - \boldsymbol{d}_i \boldsymbol{\rho}_i \boldsymbol{s}'_i$ for each $1 \le i \le l$.

Correctness. For each $1 \leq i \leq l$, we have

$$\begin{aligned}
\boldsymbol{z}_i &= \boldsymbol{b}_i + \boldsymbol{\lambda}_i(\boldsymbol{v}_i'' - \boldsymbol{u}_i''\boldsymbol{f}_i) \Rightarrow \mathrm{rk}(\boldsymbol{z}_i) \leq r_b + r_\lambda(r_v + r_u r_f). \\
\boldsymbol{c}_i\boldsymbol{\tau}_i\boldsymbol{s}_i &= \boldsymbol{c}_i\boldsymbol{\lambda}_i\boldsymbol{w}_i^{-1}(\boldsymbol{w}_i\boldsymbol{\eta}_i\boldsymbol{\mu}_i + \boldsymbol{w}_i\boldsymbol{v}_i\boldsymbol{h}) = \boldsymbol{c}_i\boldsymbol{\lambda}_i(\boldsymbol{\eta}_i\boldsymbol{\mu}_i + \boldsymbol{v}_i\boldsymbol{h}) \\
\boldsymbol{d}_i\boldsymbol{\rho}_i\boldsymbol{s}_i' &= -\boldsymbol{\lambda}_i\boldsymbol{\mu}_i\boldsymbol{e}_i^{-1}(\boldsymbol{\eta}_i' + \boldsymbol{c}_i\boldsymbol{\eta}_i)(\boldsymbol{e}_i + \boldsymbol{e}_i\boldsymbol{f}_i\boldsymbol{h}) = -\boldsymbol{\lambda}_i\boldsymbol{\mu}_i(\boldsymbol{\eta}_i' + \boldsymbol{c}_i\boldsymbol{\eta}_i)(1 + \boldsymbol{f}_i\boldsymbol{h}) \\
\boldsymbol{\gamma}_i &= \boldsymbol{a}_i + \boldsymbol{z}_i\boldsymbol{h} - \boldsymbol{c}_i\boldsymbol{\tau}_i\boldsymbol{s}_i - \boldsymbol{d}_i\boldsymbol{\rho}_i\boldsymbol{s}_i' \\
&= \boldsymbol{a}_i + \boldsymbol{b}_i\boldsymbol{h} + \boldsymbol{\lambda}_i\boldsymbol{h}(\boldsymbol{v}_i' + \boldsymbol{c}_i\boldsymbol{v}_i) - \boldsymbol{\lambda}_i\boldsymbol{f}_i\boldsymbol{h}\boldsymbol{\mu}_i(\boldsymbol{\eta}_i' + \boldsymbol{c}_i\boldsymbol{\eta}_i) \\
&\quad - \boldsymbol{c}_i\boldsymbol{\lambda}_i(\boldsymbol{\eta}_i\boldsymbol{\mu}_i + \boldsymbol{v}_i\boldsymbol{h}) + \boldsymbol{\lambda}_i\boldsymbol{\mu}_i(\boldsymbol{\eta}_i' + \boldsymbol{c}_i\boldsymbol{\eta}_i)(1 + \boldsymbol{f}_i\boldsymbol{h}) \\
&= \boldsymbol{a}_i + \boldsymbol{b}_i\boldsymbol{h} + \boldsymbol{\lambda}_i\boldsymbol{v}_i'\boldsymbol{h} + \boldsymbol{\lambda}_i\boldsymbol{\eta}_i'\boldsymbol{\mu}_i \\
\Rightarrow\quad & \boldsymbol{c} = \mathcal{H}\left(\{\boldsymbol{\gamma}_i, \boldsymbol{\tau}_i, \boldsymbol{d}_i\}_{1\leq i\leq l}, \boldsymbol{m}, \mathsf{pk}\right).
\end{aligned}$$

4 New Problems in Rank Metric Code-Based Cryptography

In this section, we introduce some existing problems (Problem 2 and 3) in rank metric code-based cryptography, which our signature scheme is based on. In addition, we define some new problems (Problem 4, 5 and 6) in rank metric code-based cryptography. Furthermore, we also discuss the hardness of these problems.

Problem 2 (Decisional Rank Syndrome Decoding (DRSD) Problem). Let H be a full rank $(n-k)\times n$ matrix over $\mathbb{F}_{q^m}$, $\boldsymbol{s} \in \mathbb{F}_{q^m}^{n-k}$, r an integer and $\boldsymbol{x} \in E_{m,n,r}$. The Decisional Rank Syndrome Decoding problem $\mathsf{DRSD}_H(q,m,n,k,r)$ is to distinguish the pair $(H, \boldsymbol{s} = \boldsymbol{x}H^T)$ from $(H, \boldsymbol{y})$ where $\boldsymbol{y} \xleftarrow{\$} \mathbb{F}_{q^m}^{n-k}$.

Remark 2. The RSD and DRSD problem is defined for random codes, but can be specialized to the families of random ideal codes. We denote RSD and DRSD problem for random ideal codes as I-RSD and I-DRSD problem respectively. Although I-RSD and I-DRSD have not been proved NP-Complete, these problems are considered hard by the research community since the best known attacks on these problems are of exponential order.

We give the following definition for advantage Adv of an adversary $\mathcal{A}$ in winning a game $\mathcal{G}$:

Definition 4. The advantage of an adversary $\mathcal{A}$ in winning a game $\mathcal{G}$, denoted as $\mathsf{Adv}(\mathcal{G})$ is defined as the probability that $\mathcal{A}$ wins the game $\mathcal{G}$.

Assumption 1 (I-RSD and I-DRSD Assumption). The Rank Syndrome Decoding for random ideal codes (I-RSD) assumption is the assumption that the advantage Adv(I-RSD) is negligible i.e., $\mathsf{Adv}(\text{I-RSD}) < \varepsilon_{\text{I-RSD}}$. The Decisional Rank Syndrome Decoding for random ideal codes (I-DRSD) assumption is the assumption that the advantage Adv(I-DRSD) is negligible i.e., $\mathsf{Adv}(\text{I-DRSD}) < \varepsilon_{\text{I-DRSD}}$.

The next problem is a problem defined in [4, Problem VI.1]:

Problem 3 (Ideal LRPC Codes Indistinguishability (I-LRPC.IND)**).** Given a polynomial $P \in \mathbb{F}_q[X]$ of degree k and a vector $\boldsymbol{h} \in \mathbb{F}_{q^m}^k$. The Ideal LRPC Codes Indistinguishability (I-LRPC.IND) problem is to distinguish whether the ideal code $\mathcal{C}$ with parity check matrix generated by $\boldsymbol{h}$ and P is a random ideal code or whether it is an ideal LRPC code of weight d. Equivalently, the I-LRPC.IND problem is to distinguish whether $\boldsymbol{h}$ is sampled uniformly at random or as $\boldsymbol{x}^{-1}\boldsymbol{y} \bmod P$ where the vectors $\boldsymbol{x}$ and $\boldsymbol{y}$ have the same support of dimension d.

By considering the arguments given in [4], we assume that solving an I-LRPC.IND problem is hard:

Assumption 2 (I-LRPC.IND **Assumption** [4]**).** The Ideal LRPC Codes Indistinguishability (I-LRPC.IND) assumption is the assumption that the advantage Adv(I-LRPC.IND) is negligible i.e., Adv(I-LRPC.IND) $< \varepsilon_{\text{I-LRPC.IND}}$.

We now introduce a new problem in rank metric code-based cryptography:

Problem 4 (Rank Support Basis Decomposition (RSBD)**).** Let $X \subset \mathbb{F}_{q^m}$ be an rd-dimensional product space such that $X = A.B$, where $A \in \mathsf{SubSp}(r, \mathbb{F}_{q^m})$ and $B \in \mathsf{SubSp}(d, \mathbb{F}_{q^m})$. Given X, the Rank Support Basis Decomposition $\mathsf{RSBD}(q, m, r, d)$ problem is to determine bases for A and B such that $X = A.B$, $\dim(A) = r$ and $\dim(B) = d$.

The following result is a prerequisite to determine a bound for solving RSBD:

Lemma 2. Let $X \subset \mathbb{F}_{q^m}$ be a product space such that $\dim(X) = rd$ and $X = A.B$, where $A = \langle a_1, \ldots, a_r\rangle \in \mathsf{SubSp}(r, \mathbb{F}_{q^m})$ and $B = \langle b_1, \ldots, b_d\rangle \in \mathsf{SubSp}(d, \mathbb{F}_{q^m})$. Suppose further that j is the integer satisfying $1 < j < d$, $rd - (j-1)(m - rd) > r$ and $rd - j(m - rd) \leq r$. Then, for all $1 \leq i \leq j$, we have

$$\dim\left(b_1^{-1}.X \cap \ldots \cap b_i^{-1}.X\right) = rd - (i-1)(m - rd) \geq r,$$
$$\dim\left(b_1^{-1}.X \cap \ldots \cap b_{j+1}^{-1}.X\right) = r.$$

Proof. Denote $B_i := b_i^{-1}X$ and $B_{1,\ldots,i} = B_1 \cap \ldots \cap B_i$. Note that for $1 \leq i \leq j$, we have $\dim(B_{1,\ldots,i} \oplus B_{i+1}) = \dim(B_{1,\ldots,i}) + \dim(B_{i+1}) - \dim(B_{1,\ldots,i} \cap B_{i+1})$.

When $i = 1$, we have $\dim(B_1) = \dim(X) = rd$, since $B_1 = b_1^{-1}X$. Therefore the statement holds for $i = 1$.

Now we proceed to $i = 2$. Since $\dim(B_1) = \dim(B_2) = rd$ and $2rd \geq m + r$, we have $\dim(B_1 \oplus B_2) = m$ and $\dim(B_1 \cap B_2) = 2rd - m = rd - (m - rd)$. The statement holds for $i = 2$.

Now we proceed to $i = 3$. Since $\dim(B_{12}) = 2rd - m$, $\dim(B_3) = rd$ and $3rd - m \geq r + m$, we have $\dim(B_{12} \oplus B_3) = m$ and $\dim(B_{12} \cap B_3) = 3rd - 2m = rd - 2(m - rd)$. The statement holds for $i = 3$.

By induction on i, we assume the statement holds for $i = j-1$. Now we consider $i = j$. By induction hypothesis, we have $\dim(B_{1,\ldots,j-1}) = (j-1)rd - (j-2)m$. Since $\dim(B_i) = rd$ and $jrd-(j-2)m \geq r+m$, we have $\dim(B_{1,\ldots,j-1} \oplus B_i) = m$ and $\dim(B_{1,\ldots,j-1} \cap B_i) = jrd-(j-1)m = rd-(j-1)(m-rd)$. The statement holds for $i = j$.

Finally, we want to show that $\dim(B_{1,\ldots,j+1}) = r$. Since $\dim(B_{1,\ldots,j}) = jrd - (j-1)m$ and $\dim(B_{j+1}) = rd$, we have $\dim(B_{1,\ldots,j}) + \dim(B_{j+1}) = (j+1)rd - (j-1)m < r+m$, $\dim(B_{1,\ldots,j} \oplus B_{j+1}) < m$ and $\dim(B_{1,\ldots,j} \cap B_{j+1}) = r$. This completes the proof for the statement. □

We consider two main approaches in solving the RSBD problem: the combinatorial approach and the algebraic approach.

Combinatorial Approach. Let $A = \langle a_1, \ldots, a_r \rangle$ and $B = \langle b_1, \ldots, b_d \rangle$, then $X = \langle a_1 b_1, \ldots, a_r b_d \rangle$. Consider a basis $\{x_1, \ldots, x_{rd}\}$ for X. We consider two different cases to recover A and B.

Case 1: Determine a basis for B. We first try to determine a basis for B, then we apply the Rank Support Recovery (RSR) algorithm [4, Algorithm 7] (refer to Appendix A) to recover A.

Let $\mathbb{j}_{m,r_1,r_2} := \left\lceil \frac{r_1 - r_2}{m - r_1} \right\rceil > 0$. Equivalently, $\mathbb{j}_{m,rd,d}$ is the integer such that

$$rd - (\mathbb{j}_{m,rd,d} - 1)(m - rd) > d, \quad rd - \mathbb{j}_{m,rd,d}(m - rd) \leq d.$$

By Lemma 2, $\mathbb{j}_{m,rd,d} + 1$ $\mathbb{F}_q$-linearly independent elements $\hat{a}_1, \ldots, \hat{a}_{\mathbb{j}_{m,rd,d}+1} \in A$ are required so that $\dim(\hat{a}_1^{-1}.X \cap \ldots \cap \hat{a}_{\mathbb{j}_{m,rd,d}+1}^{-1}.X) = d$, giving us $B = \hat{a}_1^{-1}.X \cap \ldots \cap \hat{a}_{\mathbb{j}_{m,rd,d}+1}^{-1}.X$. We sample these $\mathbb{j}_{m,rd,d}+1$ elements randomly from $\mathbb{F}_{q^m}$. The probability that the random $\hat{a}_1, \ldots, \hat{a}_{\mathbb{j}_{m,rd,d}+1}$ belong to A is $\frac{q^{(\mathbb{j}_{m,rd,d}+1)(r-\mathbb{j}_{m,rd,d}-1)}}{q^{(\mathbb{j}_{m,rd,d}+1)(m-\mathbb{j}_{m,rd,d}-1)}} = q^{-(\mathbb{j}_{m,rd,d}+1)(m-r)}$. Therefore the complexity to recover B is $O\left(q^{(\mathbb{j}_{m,rd,d}+1)(m-r)}\right)$.

Case 2: Determine a basis for A. Similar as the arguments in case 1, we first try to determine a basis for A, then we apply the RSR algorithm to recover B. The argument follows analogously as in case 1. Therefore, the complexity to recover A is $O\left(q^{(\mathbb{j}_{m,rd,r}+1)(m-d)}\right)$.

The following result summarizes the complexity to solve an RSBD problem via combinatorial approach:

Theorem 1. Let $\mathbb{j}_{m,r_1,r_2} := \left\lceil \frac{r_1 - r_2}{m - r_1} \right\rceil > 0$. Then, the complexity to solve $\mathsf{RSBD}(q, m, r, d)$ via combinatorial approach is bounded by

$$O\left(\min\left\{q^{(\mathbb{j}_{m,rd,d}+1)(m-r)}, q^{(\mathbb{j}_{m,rd,r}+1)(m-d)}\right\}\right).$$

Algebraic Approach. Since X is known, we can determine a basis $\{x_1, \ldots, x_{rd}\}$ for X. Let $A = \langle a_1, \ldots, a_r \rangle$ and $B = \langle b_1, \ldots, b_d \rangle$. For $1 \leq t \leq rd$, there exists $P \in \mathrm{GL}_{rd}(\mathbb{F}_q)$ such that

$$(x_1, \ldots, x_{rd})P = (a_1 b_1, \ldots, a_r b_d) \Leftrightarrow a_i b_j = \sum_{t=1}^{rd} p_{t,i,j} x_t$$

where $p_{t,i,j} \in \mathbb{F}_q$. Also, for all $1 \leq i \leq r$ and $1 \leq j \leq d$,

$$a_i = \sum_{t_a=1}^{m} \gamma_{t_a,i} \beta_{t_a}, \quad b_j = \sum_{t_b=1}^{m} \delta_{t_b,j} \beta_{t_b}.$$

For $1 \leq i \leq r$ and $1 \leq j \leq d$, we can rewrite $a_i b_j$ as

$$a_i b_j = \left(\sum_{t_a=1}^{m} \gamma_{t_a,i} \beta_{t_a} \right) \left(\sum_{t_b=1}^{m} \delta_{t_b,j} \beta_{t_b} \right) = \sum_{t=1}^{rd} p_{i,j,t} x_t \quad (2)$$

The above system (2) is a multivariate quadratic system with

$$\begin{aligned} \gamma_{t_a,i}, \delta_{t_b,j} &: m(r+d) \text{ quadratic unknown variables over } \mathbb{F}_q \\ p_{i,j,t} &: (rd)^2 \text{ linear unknown variables over } \mathbb{F}_q \\ a_i b_j &: mrd \text{ equations over } \mathbb{F}_q \end{aligned}$$

giving us $n' = m(r+d) + (rd)^2$ unknown variables and $m' = mrd$ equations.

We consider the best known attacks in solving multivariate quadratic system under $\mathbb{F}_q$. Note that the attack using Gröbner basis is rather inefficient for the settings in (2). The solving complexity by algebraic approach is of high exponential order.

We now give the definition for the decisional version of RSBD problem:

Problem 5 (Decisional Rank Support Basis Decomposition Problem (DRSBD)). Let $X \subset \mathbb{F}_{q^m}$ be an rd-dimensional product space such that $X = A.B$, where $A \in \mathsf{SubSp}(r, \mathbb{F}_{q^m})$ and $B \in \mathsf{SubSp}(d, \mathbb{F}_{q^m})$. The Decisional Rank Support Basis Decomposition $\mathsf{DRSBD}(q, m, r, d)$ problem is to distinguish the subspace X from the subspace X' where $X' \stackrel{\$}{\leftarrow} \mathsf{SubSp}(rd, \mathbb{F}_{q^m})$.

Next, we introduce the following problem which appears naturally when we try to prove the indistinguishability of the signatures.

Problem 6 (Decisional Advanced Rank Support Basis Decomposition Problem (DRSBD$^+$)). Let $U \stackrel{\$}{\leftarrow} \mathsf{SubSp}(w, \mathbb{F}_{q^m})$ be a random subspace and $X \subset \mathbb{F}_{q^m}$ be an rd-dimensional product space such that $U \cap X = \{0\}$ and $X = A.B$, where $A \in \mathsf{SubSp}(r, \mathbb{F}_{q^m})$ and $B \in \mathsf{SubSp}(d, \mathbb{F}_{q^m})$. The Decisional Advanced Rank Support Basis Decomposition $\mathsf{DRSBD}^+(q, m, r, d, w)$ problem is to distinguish the subspace $W = U + X$ from the subspace W' where $W' \stackrel{\$}{\leftarrow} \mathsf{SubSp}(w + rd, \mathbb{F}_{q^m})$.

It is clear that $\mathsf{DRSBD}(q, m, r, d) = \mathsf{DRSBD}^+(q, m, r, d, 0)$, i.e., every instance in DRSBD is an instance in DRSBD^+. This implies that if there exists an efficient algorithm to solve DRSBD^+, then there exists an efficient algorithm to solve DRSBD. We have the following result on the hardness of DRSBD^+ problem:

Proposition 1. Solving a $\mathsf{DRSBD}^+(q, m, r, d, w)$ problem is as hard as solving a $\mathsf{DRSBD}(q, m, r, d)$ problem.

The most efficient known combinatorial attack against the $\mathsf{RSBD}(q, m, r, d)$ is presented in Theorem 1, which is of exponential order. Moreover, solving the RSBD problem via the algebraic approach requires high exponential complexity, as argued above via the modeling (2). By Proposition 1 and the arguments above, we make the following assumption on DRSBD^+ problem:

Assumption 3 (DRSBD^+ Assumption). The Decisional Advanced Rank Support Basis Decomposition (DRSBD^+) assumption is the assumption that the advantage $\mathsf{Adv}(\mathsf{DRSBD}^+)$ is negligible i.e., $\mathsf{Adv}(\mathsf{DRSBD}^+) < \varepsilon_{\mathsf{DRSBD}^+}$.

5 Provable Security and Practical Security

One of the desired security models for digital signature schemes is existential unforgeability under an adaptive chosen message attack (EUF-CMA). This is defined by a security game which is interacting between a challenger and an adversary $\mathcal{A}$. If $\mathcal{A}$ has access to a signature oracle, it cannot produce a valid signature for a new message with non-negligible probability. The security game $\mathbf{Exp}^{\mathsf{euf}}_{\mathcal{S},\mathcal{A}}(\lambda)$ is described as follows:

Setup: Given a security parameter λ, the challenger first runs the Key.Gen algorithm. The public key pk is sent to the adversary $\mathcal{A}$ and the secret key sk is kept to the challenger. The challenger sets $\mathcal{SM} = \emptyset$.

Signature Queries: The adversary $\mathcal{A}$ issues signature queries $m_1, \ldots, m_N$ to the challenger. For each query m_i, the challenger responds by running Sign to generate the signature σ_i of m_i and sending σ_i to $\mathcal{A}$. The challenger also updates $\mathcal{SM} = \mathcal{SM} \cup \{m_i\}$. These queries may be asked adaptively so that each query m_i may depend on the replies to $m_1, \ldots, m_{i-1}$.

Output: The adversary $\mathcal{A}$ outputs a pair $(m, \boldsymbol{\sigma})$. The adversary wins if $\boldsymbol{\sigma}$ is a valid signature of m according to Verify and $m \notin \mathcal{SM}$.

The probability of success against this game is denoted by

$$\mathsf{Succ}^{\mathsf{euf}}_{\mathcal{S},\mathcal{A}}(\lambda) = \mathsf{Pr}\left(\mathbf{Exp}^{\mathsf{euf}}_{\mathcal{S},\mathcal{A}}(\lambda) = 1\right), \quad \mathsf{Succ}^{\mathsf{euf}}_{\mathcal{S}}(\lambda, t) = \max_{\mathcal{A} \leq t} \mathsf{Succ}^{\mathsf{euf}}_{\mathcal{S},\mathcal{A}}(\lambda).$$

Definition 5 (EUF-CMA [20]). A signature scheme is (t, N, ϵ)-existentially unforgeable under an adaptive chosen message attack (EUF-CMA) if for any probabilistic t-polynomial time, the adversary $\mathcal{A}$ making at most N signature queries has the advantage less than ϵ, i.e., $\mathsf{Succ}^{\mathsf{euf}}_{\mathcal{S},\mathcal{A}}(\lambda) < \epsilon$.

Theorem 2 (EUF-CMA **Security**). Under the I-LRPC.IND, DRSBD$^+$, I-DRSD and I-RSD assumptions, our signature scheme is secure under the EUF-CMA model in the Random Oracle Model.

Proof. Let $\mathcal{A}$ be an adversary who can break our signature scheme, then there exists an attacker $\mathfrak{A}$ who can break the I-LRPC.IND, DRSBD$^+$, I-DRSD and I-RSD assumptions. To prove the security of the scheme, we are using a sequence of games, $\mathcal{G}_j$. Let $r_z = r_b + r_\lambda(r_v + r_u r_f)$. For $j \geq 0$, denote $\mathsf{Adv}(\mathcal{G}_j)$ as the advantage of $\mathcal{A}$ in game $\mathcal{G}_j$.

Game $\mathcal{G}_0$: This is the real EUF-CMA game for $\mathcal{S}$. The adversary has access to the signature oracle Sign to obtain valid signatures. Then

$$\mathsf{Adv}(\mathcal{G}_0) = \mathsf{Succ}^{\mathsf{euf}}_{\mathcal{S},\mathcal{A}}(\lambda).$$

Game $\mathcal{G}_1$: In $\mathcal{G}_0$, for $1 \leq i \leq l$, we have $\boldsymbol{\tau}_i = \boldsymbol{\lambda}_i \boldsymbol{w}_i^{-1}$ and $\boldsymbol{d}_i = -\boldsymbol{\lambda}_i \boldsymbol{\mu}_i \boldsymbol{e}_i^{-1}$. We now replace $\boldsymbol{\tau}_i$ by a vector $\boldsymbol{\tau}'_i \overset{\$}{\leftarrow} E_{m,k,k}$. This corresponds to an instance of the I-LRPC.IND problem. Similarly, we replace $\boldsymbol{d}_i$ by a vector $\boldsymbol{d}'_i \overset{\$}{\leftarrow} E_{m,k,k}$, which corresponds to an instance of the I-LRPC.IND problem. Therefore, we have

$$|\mathsf{Adv}(\mathcal{G}_1) - \mathsf{Adv}(\mathcal{G}_0)| \leq 2l \times \mathsf{Adv}(\text{I-LRPC.IND}).$$

Game $\mathcal{G}_2$: For each $1 \leq i \leq l$, we replace $\boldsymbol{a}_i$ by a vector $\boldsymbol{a}'_i \overset{\$}{\leftarrow} A_i^k$ and replace $\boldsymbol{z}_i$ by a vector $\boldsymbol{z}'_i \overset{\$}{\leftarrow} Z_i^k$ where $A_i = \mathrm{supp}(\boldsymbol{a}_i)$ and $Z_i = \mathrm{supp}(\boldsymbol{z}_i)$. Then, we sample $\boldsymbol{c}'_i$, $\boldsymbol{\tau}'_i$, $\boldsymbol{d}'_i$ and $\boldsymbol{\rho}'_i$ uniformly. Set $\boldsymbol{\gamma}'_i = (\boldsymbol{a}'_i + \boldsymbol{z}'_i\boldsymbol{h}) - \boldsymbol{c}'_i\boldsymbol{\tau}'\boldsymbol{s}_i - \boldsymbol{d}'_i\boldsymbol{\rho}'_i\boldsymbol{s}'_i$ and use the random oracle to set $\boldsymbol{c} = \mathcal{H}\left(\{\boldsymbol{\gamma}'_i, \boldsymbol{\tau}'_i, \boldsymbol{d}'_i\}_{1 \leq i \leq l}, \boldsymbol{m}, \mathsf{pk}\right)$. Note that the distribution of $\boldsymbol{\gamma}'_i$ is the uniform distribution over $\mathbb{F}_{q^m}^{n-k}$, and each $\boldsymbol{\gamma}'_i$ is independent from each other.

Now, consider $B_i \in \mathsf{SubSp}(r_b, \mathbb{F}_{q^m})$, $\mathcal{U}_i \in \mathsf{SubSp}(r_u, \mathbb{F}_{q^m})$, $\mathcal{V}_i \in \mathsf{SubSp}(r_v, \mathbb{F}_{q^m})$ and $\Lambda_i = \mathrm{supp}(\boldsymbol{\lambda}_i) \in \mathsf{SubSp}(r_\lambda, \mathbb{F}_{q^m})$. Suppose that $\boldsymbol{a}'_i \overset{\$}{\leftarrow} (A_i + \Lambda_i.\mathcal{U}_i)^k$ and $\boldsymbol{b}'_i \overset{\$}{\leftarrow} (B_i + \Lambda_i.\mathcal{V}_i)^k$. Let $\Delta(\mathcal{X}_i, \mathcal{X}'_i)$ be the statistical distance between $\mathcal{X}_i$ and $\mathcal{X}'_i$, where $\mathcal{X}_i$ is the distribution of $\boldsymbol{x}_i = \boldsymbol{a}'_i + \boldsymbol{b}'_i\boldsymbol{h}$ and $\mathcal{X}'_i$ is the uniform distribution over $\mathbb{F}_{q^m}^k$. Let Φ_i be a family of functions defined by

$$\begin{aligned}\Phi_i = \{\phi_{\boldsymbol{h}} : (A_i + \Lambda_i.\mathcal{U}_i)^k \times (B_i + \Lambda_i.\mathcal{V}_i)^k \to \mathbb{F}_{q^m}^k \\ \text{such that } \phi_{\boldsymbol{h}}(\boldsymbol{a}'_i, \boldsymbol{b}'_i) = \boldsymbol{a}'_i + \boldsymbol{b}'_i\boldsymbol{h} = \boldsymbol{x}_i\}.\end{aligned}$$

Since $\boldsymbol{h}$ is chosen uniformly at random, then Φ_i is a pairwise independent family of functions. The number of choices for $(\boldsymbol{a}'_i, \boldsymbol{b}'_i)$ depends on A_i, B_i, Λ_i and the coordinates of $(\boldsymbol{a}'_i, \boldsymbol{b}'_i)$. Overall, the entropy of $(\boldsymbol{a}'_i, \boldsymbol{b}'_i)$ is

$$\Theta\left(\begin{bmatrix} m \\ r_a \end{bmatrix}_q \begin{bmatrix} m \\ r_b \end{bmatrix}_q \begin{bmatrix} m \\ r_\lambda \end{bmatrix}_q q^{(r_a + r_b + r_\lambda(r_u + r_v))k}\right) = 2^{\delta(\Phi)\log q + O(1)}$$

where $\delta(\Phi) = r_a(m - r_a) + r_b(m - r_b) + r_\lambda(m - r_\lambda) + (r_a + r_b + r_\lambda(r_u + r_v))k$. Since $\mathrm{rk}(\boldsymbol{a}'_i, \boldsymbol{b}'_i) = r_a + r_b + r_\lambda(r_u + r_v) > d_{\mathsf{RGV}}$, any vector of $\mathbb{F}_{q^m}^k$ can be reached, giving us the entropy of $\boldsymbol{x}_i$ is equal to $2^{km \log q}$. By Leftover Hash Lemma [22], we have $\Delta(\mathcal{X}_i, \mathcal{X}_i') \leq \frac{\varepsilon}{2}$ where $\varepsilon = 2^{\frac{[km-\delta(\Phi_i)]\log q}{2}} + O(1)$ and $\delta(\Phi_i) = r_a(m-r_a) + r_b(m-r_b) + r_\lambda(m-r_\lambda) + (r_a + r_b + r_\lambda(r_u + r_v))k$. Therefore, we have

$$|\mathsf{Adv}(\mathcal{G}_2) - \mathsf{Adv}(\mathcal{G}_1)| \leq l \times \varepsilon.$$

Game $\mathcal{G}_3$: For $1 \leq i \leq l$, we replace $\boldsymbol{a}_i$ by a vector $\boldsymbol{a}_i' \xleftarrow{\$} E_{m,k,r_a}$ and replace $\boldsymbol{z}_i$ by a vector $\boldsymbol{z}_i' \xleftarrow{\$} E_{m,k,r_z}$. Then, we sample $\boldsymbol{c}_i'$ uniformly. We set $\gamma_i' = (\boldsymbol{a}_i' + \boldsymbol{z}_i'\boldsymbol{h}) - \boldsymbol{c}_i'\boldsymbol{\tau}_i'\boldsymbol{s}_i - \boldsymbol{d}_i'\boldsymbol{\rho}_i'\boldsymbol{s}_i'$ and use the random oracle to set $\boldsymbol{c} = \mathcal{H}\left(\{\gamma_i', \boldsymbol{\tau}_i', \boldsymbol{d}_i'\}_{1\leq i\leq l}, \boldsymbol{m}, \mathsf{pk}\right)$. This corresponds to an instance of the DRSBD^+ problem. Therefore, we have

$$|\mathsf{Adv}(\mathcal{G}_3) - \mathsf{Adv}(\mathcal{G}_2)| \leq l \times \mathsf{Adv}(\mathsf{DRSBD}^+).$$

Game $\mathcal{G}_4$: We now pick random $\boldsymbol{s}_i, \boldsymbol{s}_i' \xleftarrow{\$} \mathbb{F}_{q^m}^k$ and proceed the steps as before. The difference between $\mathcal{G}_4$ and $\mathcal{G}_3$ resides in the public key pk. Therefore, we have

$$|\mathsf{Adv}(\mathcal{G}_4) - \mathsf{Adv}(\mathcal{G}_3)| \leq 2l \times \mathsf{Adv}(\mathsf{I\text{-}DRSD}).$$

At this step, everything we send to the adversary is random and independent from any sk. Hence, the security of our scheme is reduced to the case where no signature is given to the attacker. If $\mathcal{A}$ can compute a valid signature after game $\mathcal{G}_3$, then the challenger can compute a solution $(\boldsymbol{a}_i, \boldsymbol{z}_i)$ of the instance $(\boldsymbol{h}, \gamma_i + \boldsymbol{c}_i\boldsymbol{\tau}_i\boldsymbol{s}_i + \boldsymbol{d}_i\boldsymbol{\rho}_i\boldsymbol{s}_i', r_a + r_z)$ as an $\mathsf{I\text{-}RSD}$ problem. Therefore, we have

$$|\mathsf{Adv}(\mathcal{G}_4)| = l \times \varepsilon_{\mathsf{I\text{-}RSD}}.$$

For each $1 \leq i \leq 4$, we have $|\mathsf{Adv}(\mathcal{G}_{i-1})| - |\mathsf{Adv}(\mathcal{G}_i)| \leq |\mathsf{Adv}(\mathcal{G}_{i-1}) - \mathsf{Adv}(\mathcal{G}_i)|$. Therefore

$$\begin{aligned}
|\mathsf{Adv}(\mathcal{G}_0)| - |\mathsf{Adv}(\mathcal{G}_4)| &= \sum_{i=1}^{4} |\mathsf{Adv}(\mathcal{G}_i)| - |\mathsf{Adv}(\mathcal{G}_{i-1})| \\
&\leq \sum_{i=1}^{4} |\mathsf{Adv}(\mathcal{G}_i) - \mathsf{Adv}(\mathcal{G}_{i-1})| \\
&\leq l\left(2\varepsilon_{\mathsf{I\text{-}LRPC.IND}} + \varepsilon + \varepsilon_{\mathsf{DRSBD}^+} + 2\varepsilon_{\mathsf{I\text{-}DRSD}}\right) \\
|\mathsf{Adv}(\mathcal{G}_0)| &\leq l\left(2\varepsilon_{\mathsf{I\text{-}LRPC.IND}} + \varepsilon + \varepsilon_{\mathsf{DRSBD}^+} + 2\varepsilon_{\mathsf{I\text{-}DRSD}}\right) + |\mathsf{Adv}(\mathcal{G}_4)| \\
\Rightarrow \quad \mathsf{Succ}_{\mathcal{S},\mathcal{A}}^{\mathsf{euf}}(\lambda) &\leq l\left(2\varepsilon_{\mathsf{I\text{-}LRPC.IND}} + \varepsilon + \varepsilon_{\mathsf{DRSBD}^+} + 2\varepsilon_{\mathsf{I\text{-}DRSD}} + \varepsilon_{\mathsf{I\text{-}RSD}}\right).
\end{aligned}$$

□

In the rest of this section, we analyze the structural security of our signature scheme. Let $r_z = r_b + r_\lambda(r_v + r_u r_f)$. Denote $\mathcal{O}_{\mathsf{RSD}}$ as the complexity to solve an RSD problem.

1. **Recover $\boldsymbol{w}_{uv} = (\boldsymbol{w}_i\boldsymbol{\eta}_i\boldsymbol{\mu}_i, \boldsymbol{w}_i\boldsymbol{v}_i)$ from $\boldsymbol{s}_i$.** Note that $\text{rk}(\boldsymbol{w}_{uv}) = r_w(r_u + r_v)$. If an adversary can solve an $\mathsf{RSD}\,(q, m, 2k, k, r_w(r_u + r_v))$ problem, then he can recover the vector $\boldsymbol{w}_{uv}$ from $\boldsymbol{s}_i = \boldsymbol{w}_i\boldsymbol{\mu}_i\boldsymbol{\eta}_i + \boldsymbol{w}_i\boldsymbol{v}_i\boldsymbol{h}$.
2. **Recover $\boldsymbol{e}_f = (\boldsymbol{e}_i, \boldsymbol{e}_i\boldsymbol{f}_i)$ from $\boldsymbol{s}'_i$.** Note that $\text{rk}(\boldsymbol{e}_f) = r_e(1+r_f)$. If an adversary can solve an $\mathsf{RSD}\,(q, m, 2k, k, r_e(1 + r_f))$ problem, then he can recover the vector $\boldsymbol{e}_f$ from $\boldsymbol{s}'_i = \boldsymbol{e}_i + \boldsymbol{e}_i\boldsymbol{f}_i\boldsymbol{h}$.
3. **For $1 \le i \le l$, consider $(\boldsymbol{a}_i, \boldsymbol{z}_i, \boldsymbol{c}_i, \boldsymbol{\tau}_i, \boldsymbol{d}_i, \boldsymbol{\rho}_i)$:**
   a. **Recover $\boldsymbol{w}_\lambda = (-\boldsymbol{\lambda}_i, \boldsymbol{w}_i)$.** Since $\boldsymbol{\tau}_i = \boldsymbol{\lambda}_i\boldsymbol{w}_i^{-1}$, we have $-\boldsymbol{\lambda}_i + \boldsymbol{w}_i\boldsymbol{\tau}_i = \boldsymbol{0}$. Note that $\text{rk}(\boldsymbol{w}_\lambda) = r_w + r_\lambda$. Suppose an adversary can solve an $\mathsf{RSD}(q, m, 2k, k, r_w + r_\lambda)$ problem, then he can recover the vector $\boldsymbol{w}_\lambda$ with complexity $q^{-k}\mathcal{O}_{\mathsf{RSD}}(q, m, 2k, k, r_w + r_\lambda)$.
   b. **Recover $\boldsymbol{\lambda}_{ue} = (\boldsymbol{\lambda}_i\boldsymbol{\mu}_i, \boldsymbol{e}_i)$.** Since $\boldsymbol{d}_i = -\boldsymbol{\lambda}_i\boldsymbol{\mu}_i\boldsymbol{e}_i^{-1}$, we have $\boldsymbol{\lambda}_i\boldsymbol{\mu}_i + \boldsymbol{e}_i\boldsymbol{d}_i = \boldsymbol{0}$. Note that $\text{rk}(\boldsymbol{\lambda}_{uv}) = r_\lambda r_u + r_e$. Suppose an adversary can solve an $\mathsf{RSD}(q, m, 2k, k, r_\lambda r_u + r_e)$ problem, then he can recover the vector $\boldsymbol{\lambda}_{ue}$ with complexity $q^{-k}\mathcal{O}_{\mathsf{RSD}}(q, m, 2k, k, r_\lambda r_u + r_e)$.
   c. **Recover $\boldsymbol{e}_{wu} = (\boldsymbol{e}_i, \boldsymbol{w}_i\boldsymbol{\mu}_i)$.** Note that $\text{rk}(\boldsymbol{e}_{wu}) = r_e + r_w r_u$. Since $\boldsymbol{\tau}_i = \boldsymbol{\lambda}_i\boldsymbol{w}_i^{-1}$ and $\boldsymbol{d}_i = -\boldsymbol{\lambda}_i\boldsymbol{\mu}_i\boldsymbol{e}_i^{-1}$, we have $\boldsymbol{\lambda}_i = \boldsymbol{\tau}_i\boldsymbol{w}_i = -\boldsymbol{d}_i\boldsymbol{e}_i(\boldsymbol{\mu}_i)^{-1} \Rightarrow \boldsymbol{e}_i\boldsymbol{d}_i + \boldsymbol{w}_i\boldsymbol{\mu}_i\boldsymbol{\tau}_i = \boldsymbol{0}$. Suppose an adversary can solve an $\mathsf{RSD}(q, m, 2k, k, r_e + r_w r_u)$ problem, then he can recover the vector $\boldsymbol{e}_{wu}$ with complexity $q^{-k}\mathcal{O}_{\mathsf{RSD}}(q, m, 2k, k, r_e + r_w r_u)$.
   d. **LTP Attack using available basis in $\boldsymbol{z}_i$.** Recall that $\boldsymbol{z}_i = \boldsymbol{b}_i + \boldsymbol{\lambda}_i(\boldsymbol{v}''_i - \boldsymbol{u}''_i\boldsymbol{f}_i)$, therefore $\text{supp}(\boldsymbol{z}_i) = \text{supp}\,(\boldsymbol{b}_i, \boldsymbol{\lambda}_i\boldsymbol{v}_i, \boldsymbol{\lambda}_i\boldsymbol{\mu}_i\boldsymbol{f}_i)$. We consider to apply LTP Attack on the following vectors:
      i. On $\gamma_i = \boldsymbol{a}_i + \boldsymbol{\lambda}_i\boldsymbol{\eta}'_i\boldsymbol{\mu}_i + (\boldsymbol{b}_i + \boldsymbol{\lambda}_i\boldsymbol{v}'_i)\boldsymbol{h}$: the bases for the support $\text{supp}(\boldsymbol{a}_i, \boldsymbol{\lambda}_i\boldsymbol{\eta}'_i\boldsymbol{\mu}_i)$ and $\text{supp}(\boldsymbol{b}_i, \boldsymbol{\lambda}_i\boldsymbol{v}'_i)$ are required for LTP Attack. We now aim to recover a basis that contains $\text{supp}(\boldsymbol{\lambda}_i\boldsymbol{\mu}_i)$. Assume that we know a basis for $\text{supp}(\boldsymbol{\lambda}_i\boldsymbol{\mu}_i\boldsymbol{f}_i)$. Then to determine $\text{supp}(\boldsymbol{\lambda}_i\boldsymbol{\mu}_i)$ is an instance of the $\mathsf{RSBD}(q, m, r_\lambda r_u, r_f)$ problem. On the other hand, we have $\text{supp}(\boldsymbol{b}_i, \boldsymbol{\lambda}_i\boldsymbol{v}_i) \subset \text{supp}(\boldsymbol{z}_i)$. If $(r_a + r_\lambda r_u + r_z)k < mk$, then we can solve for support matrices of $\boldsymbol{a}_i + \boldsymbol{\lambda}_i\boldsymbol{\eta}'_i\boldsymbol{\mu}_i$ and $\boldsymbol{b}_i + \boldsymbol{\lambda}_i\boldsymbol{v}'_i$. Therefore, the solving complexity is at least the complexity to solve an $\mathsf{RSBD}(q, m, r_\lambda r_u, r_f)$.
      ii. On $\boldsymbol{s}_i = \boldsymbol{w}_i(\boldsymbol{\eta}_i\boldsymbol{\mu}_i + \boldsymbol{v}_i\boldsymbol{h})$: the bases for $\text{supp}(\boldsymbol{w}_i\boldsymbol{\mu}_i)$ and $\text{supp}(\boldsymbol{w}_i\boldsymbol{v}_i)$ are required for LTP Attack. One could compute $\boldsymbol{\tau}_i^{-1}\boldsymbol{z}_i = \boldsymbol{\tau}_i^{-1}\boldsymbol{b}_i + \boldsymbol{w}_i(\boldsymbol{v}''_i - \boldsymbol{u}''_i\boldsymbol{f}_i)$. Since $\text{rk}(\boldsymbol{\tau}_i^{-1}\boldsymbol{z}_i) = \text{rk}(\boldsymbol{\tau}_i^{-1}\boldsymbol{b}_i) = k$, then the support $\text{supp}(\boldsymbol{w}_i\boldsymbol{\mu}_i) \not\subset \text{supp}(\boldsymbol{\tau}_i^{-1}\boldsymbol{z}_i)$. The support $\text{supp}(\boldsymbol{w}_i\boldsymbol{\mu}_i, \boldsymbol{w}_i\boldsymbol{v}_i)$ is not available for LTP Attack to work.
      iii. On $\boldsymbol{\tau}_i\boldsymbol{s}_i = \boldsymbol{\lambda}_i(\boldsymbol{\eta}_i\boldsymbol{\mu}_i + \boldsymbol{v}_i\boldsymbol{h})$: the bases for $\text{supp}(\boldsymbol{\lambda}_i\boldsymbol{\mu}_i)$ and $\text{supp}(\boldsymbol{\lambda}_i\boldsymbol{v}_i)$ are required for the LTP Attack. We now aim to recover a basis that contains $\text{supp}(\boldsymbol{\lambda}_i\boldsymbol{\mu}_i)$. Assume that we know a basis for $\text{supp}(\boldsymbol{\lambda}_i\boldsymbol{\mu}_i\boldsymbol{f}_i)$. Then to determine a basis for $\text{supp}(\boldsymbol{\lambda}_i\boldsymbol{\mu}_i)$ is an instance of the $\mathsf{RSBD}(q, m, r_\lambda r_u, r_f)$ problem. On the other hand, $\text{supp}(\boldsymbol{\lambda}_i\boldsymbol{\mu}_i) \subset \text{supp}(\boldsymbol{z}_i)$. If $(r_\lambda r_u + r_z)k < mk$, then we can solve for support matrices of $\boldsymbol{\lambda}_i\boldsymbol{\eta}_i\boldsymbol{\mu}_i$ and $\boldsymbol{\lambda}_i\boldsymbol{v}_i$. Therefore, solving for $(\boldsymbol{\lambda}_i\boldsymbol{\mu}_i\boldsymbol{\eta}_i, \boldsymbol{\lambda}_i\boldsymbol{v}_i)$ is at least of the complexity to solve $\mathsf{RSBD}(q, m, r_\lambda r_u, r_f)$.

iv. On $\boldsymbol{s}_i' = \boldsymbol{e}_i + \boldsymbol{e}_i\boldsymbol{f}_i\boldsymbol{h}$: the basis for $\text{supp}(\boldsymbol{e}_i, \boldsymbol{e}_i\boldsymbol{f}_i)$ is required for LTP Attack. However, $\text{supp}(\boldsymbol{e}_i, \boldsymbol{e}_i\boldsymbol{f}_i)$ is not available for LTP Attack.

v. On $\boldsymbol{d}_i\boldsymbol{s}_i' = -\boldsymbol{\lambda}_i\boldsymbol{\mu}_i - \boldsymbol{\lambda}_i\boldsymbol{\mu}_i\boldsymbol{f}_i\boldsymbol{h}$: we require a basis of $\text{supp}(\boldsymbol{\lambda}_i\boldsymbol{\mu}_i, \boldsymbol{\lambda}_i\boldsymbol{\mu}_i\boldsymbol{f}_i)$ for the LTP Attack. We now aim to recover a basis that contains $\text{supp}(\boldsymbol{\lambda}_i\boldsymbol{\mu}_i)$. Assume that we know a basis for $\text{supp}(\boldsymbol{\lambda}_i\boldsymbol{\mu}_i\boldsymbol{f}_i)$. Similarly, to determine a basis of the support $\text{supp}(\boldsymbol{\lambda}_i\boldsymbol{\mu}_i)$ is an instance of the $\mathsf{RSBD}(q, m, r_\lambda r_u, r_f)$ problem. Moreover, we have the fact that $\text{supp}(\boldsymbol{\lambda}_i\boldsymbol{\mu}_i\boldsymbol{f}_i) \subset \text{supp}(\boldsymbol{z}_i)$. If $(r_\lambda r_u + r_z)k < mk$, then we can solve for support matrices of $\boldsymbol{\lambda}_i\boldsymbol{\mu}_i$ and $\boldsymbol{\lambda}_i\boldsymbol{\mu}_i\boldsymbol{f}_i$. Therefore, the complexity to solve for the vector $(\boldsymbol{\lambda}_i\boldsymbol{\mu}_i, \boldsymbol{\lambda}_i\boldsymbol{\mu}_i\boldsymbol{f}_i)$ is at least the complexity to solve an $\mathsf{RSBD}(q, m, r_\lambda r_u, r_f)$.

4. **Consider two signatures, $\boldsymbol{\sigma}_j = \left(\{\boldsymbol{c}_{ji}, \boldsymbol{a}_{ji}, \boldsymbol{z}_{ji}, \boldsymbol{\tau}_{ji}, \boldsymbol{d}_{ji}, \boldsymbol{\rho}_{ji}\}_{1\leq i\leq l}\right)$ for** $j = 1, 2$. For a fixed $1 \leq i \leq l$, we have $\boldsymbol{d}_{ji} = -\boldsymbol{\lambda}_{ji}\boldsymbol{\mu}_i\boldsymbol{e}_i^{-1}$. Then, $-\boldsymbol{e}_i = \boldsymbol{\lambda}_{1i}\boldsymbol{\mu}_i\boldsymbol{d}_{1i}^{-1} = \boldsymbol{\lambda}_{2i}\boldsymbol{\mu}_i\boldsymbol{d}_{2i}^{-1}$. This implies that $\boldsymbol{0} = \boldsymbol{\lambda}_{2i}\boldsymbol{\mu}_i\boldsymbol{d}_{2i}^{-1} - \boldsymbol{\lambda}_{1i}\boldsymbol{\mu}_i\boldsymbol{d}_{1i}^{-1} = \boldsymbol{\mu}_i\left(\boldsymbol{\lambda}_{2i}\boldsymbol{d}_{2i}^{-1} - \boldsymbol{\lambda}_{1i}\boldsymbol{d}_{1i}^{-1}\right)$, and thus $\boldsymbol{0} = \boldsymbol{\lambda}_{2i}\boldsymbol{d}_{1i} - \boldsymbol{\lambda}_{1i}\boldsymbol{d}_{2i}$. Note that $\text{rk}(\boldsymbol{\lambda}_{2i}, \boldsymbol{\lambda}_{1i}) = 2r_\lambda$. Suppose that an adversary can solve an $\mathsf{RSD}(q, m, 2k, k, 2r_\lambda)$ problem, then he can recover the secret ephemeral vector $(\boldsymbol{\lambda}_{2i}, \boldsymbol{\lambda}_{1i})$ with complexity of $q^{-k}\mathcal{O}_{\mathsf{RSD}}(q, m, 2k, k, 2r_\lambda)$.

5. **Reuse $\boldsymbol{a}_i$ and $\boldsymbol{z}_i$ to forge a signature with a new message $\boldsymbol{m}'$.** Suppose a forger collected a signature $\boldsymbol{\sigma} = \left(\{\boldsymbol{c}_i, \boldsymbol{a}_i, \boldsymbol{z}_i, \boldsymbol{\tau}_i, \boldsymbol{d}_i, \boldsymbol{\rho}_i\}_{1\leq i\leq l}\right)$. Then the forger reuse γ_i, $\boldsymbol{a}_i$, $\boldsymbol{z}_i$, $\boldsymbol{\tau}_i$ and $\boldsymbol{d}_i$ to sign a new signature $\boldsymbol{m}'$. In particular, the forger computes $\boldsymbol{c}' = \mathcal{H}\left(\{\gamma_i, \boldsymbol{\tau}_i', \boldsymbol{d}_i'\}_{1\leq i\leq l}, \mathsf{pk}, \boldsymbol{m}'\right)$, where $\boldsymbol{\tau}_i'$ and $\boldsymbol{d}_i'$ are of the forger's own choices. To ensure the signature is valid, we require

$$\begin{aligned}\gamma_i &= \boldsymbol{a}_i + \boldsymbol{z}_i\boldsymbol{h} - \boldsymbol{c}_i\boldsymbol{\tau}_i\boldsymbol{s}_i - \boldsymbol{\rho}_i\boldsymbol{d}_i\boldsymbol{s}' \\ &= \boldsymbol{a}_i + \boldsymbol{z}_i\boldsymbol{h} - \boldsymbol{c}_i\boldsymbol{\tau}_i\boldsymbol{s}_i - \boldsymbol{\rho}_i\boldsymbol{d}_i\boldsymbol{s}' + \boldsymbol{c}_i'\boldsymbol{\tau}_i'\boldsymbol{s}_i + \boldsymbol{\rho}_i'\boldsymbol{d}_i'\boldsymbol{s}' - \boldsymbol{c}_i'\boldsymbol{\tau}_i'\boldsymbol{s}_i - \boldsymbol{\rho}_i'\boldsymbol{d}_i'\boldsymbol{s}' \\ &= \boldsymbol{a}_i + \boldsymbol{z}_i\boldsymbol{h} - \boldsymbol{c}_i'\boldsymbol{\tau}_i'\boldsymbol{s}_i - \boldsymbol{\rho}_i'\boldsymbol{d}_i'\boldsymbol{s}'.\end{aligned}$$

This implies that we require $\boldsymbol{c}_i'\boldsymbol{\tau}_i'\boldsymbol{s}_i + \boldsymbol{\rho}_i'\boldsymbol{d}_i'\boldsymbol{s}' - \boldsymbol{c}_i\boldsymbol{\tau}_i\boldsymbol{s}_i - \boldsymbol{\rho}_i\boldsymbol{d}_i\boldsymbol{s}' = \boldsymbol{0}$, or equivalently, $\boldsymbol{\rho}_i' = (\boldsymbol{\rho}_i\boldsymbol{d}_i + (\boldsymbol{c}_i\boldsymbol{\tau}_i - \boldsymbol{c}_i'\boldsymbol{\tau}_i')\boldsymbol{s}_i(\boldsymbol{s}_i')^{-1})(\boldsymbol{d}_i')^{-1}$. However, $\boldsymbol{\rho}_i' \notin \mathbb{F}_q^k$, since $\text{rk}(\boldsymbol{\rho}_i') > 1$. Therefore, this forgery method would fail.

6. **Suppose that $\boldsymbol{\eta}_i'$ and $\boldsymbol{\eta}_i$ are available from $\boldsymbol{\rho}_i$.** Suppose that an adversary is able to solve an $\mathsf{RSD}(q, m, 2k, k, r_w(r_u + r_v))$ problem on syndrome $\boldsymbol{s}_i = \boldsymbol{w}_i\boldsymbol{\eta}_i\boldsymbol{\mu}_i + \boldsymbol{w}_i\boldsymbol{v}_i\boldsymbol{h}$ associated with parity check matrix generated by $(\boldsymbol{\eta}_i, \boldsymbol{h})$, then he can recover the secret vector $(\boldsymbol{w}_i\boldsymbol{\mu}_i, \boldsymbol{w}_i\boldsymbol{v}_i)$ with complexity of $q^{-k}\mathcal{O}_{\mathsf{RSD}}(q, m, 2k, k, r_w(r_u + r_v))$. The arguments for solving $\boldsymbol{\tau}_i\boldsymbol{s}_i = \boldsymbol{\lambda}_i\boldsymbol{\mu}_i\boldsymbol{\eta}_i + \boldsymbol{\lambda}_i\boldsymbol{v}_i\boldsymbol{h}$ and $\gamma_i = (\boldsymbol{a}_i(\boldsymbol{\eta}_i')^{-1} + \boldsymbol{\lambda}_i\boldsymbol{\mu}_i)\boldsymbol{\eta}_i' + (\boldsymbol{b}_i + \boldsymbol{\lambda}_i\boldsymbol{v}_i)\boldsymbol{h}$ follow similarly.

6 Suggested Parameters

In this section, we discuss the rationales for the choices of our parameters.

We consider $q = 2$, $r_\lambda = r_w$ and $r_u = r_v = r_e = r_f$. For m, we choose m to be a prime so that $\mathbb{F}_{q^m}$ has no subfield, as it is a common choice in rank metric based cryptosystems. Moreover, notice that we require $r_z = r_b + r_\lambda(r_u + r_v r_f) < k$, which gives us k to be the order of at least $O(r_\lambda r_u^2)$. Suppose we choose $r_\lambda > r_u$, then we require r_u to have order of $O\left(k^{\frac{1}{3}}\right)$.

Furthermore, the collision-resistant hash function $\mathcal{H}$ has output with length kl-bit in $\mathbb{F}_q$. To achieve an $\mathsf{Sec_{CL}}$-classical security level against the collision attack, the length of output for the hash function has to be at least $2 \times \mathsf{Sec_{CL}}$ bits. In the quantum setting, the length of output for the has function has to be at least $3 \times \mathsf{Sec_{PQ}}$ for an $\mathsf{Sec_{PQ}}$-post-quantum security level. Thus, we set l to be $\left\lceil \frac{2\mathsf{Sec_{CL}}}{k\log_2(q)} \right\rceil$ and $\left\lceil \frac{3\mathsf{Sec_{PQ}}}{k\log_2(q)} \right\rceil$ respectively for $\mathsf{Sec_{CL}}$-classical and $\mathsf{Sec_{PQ}}$-post-quantum security.

Note that $\boldsymbol{h}$ can be recovered from a seed of 256 bits since $\boldsymbol{h}$ is a random vector. Moreover, we can apply Lemma 1 to represent the secret key and signature in the form of support basis and support matrix. Let $b_{i,m,k} = \min\{i(m+k), km\}$ and $r_z = r_b + r_\lambda(r_v + r_u r_f)$. The public key size ($\mathsf{size_{pk}}$), secret key size ($\mathsf{size_{sk}}$) and signature size ($\mathsf{size_\sigma}$) are in bytes (Table 3):

$$\begin{aligned}
\mathsf{size_{pk}} &= \frac{2lkm}{8}\log_2(q) + 32,\\
\mathsf{size_{sk}} &= \frac{l}{8}\left(b_{r_w,m,k} + b_{r_u,m,k} + b_{r_v,m,k} + b_{r_e,m,k} + b_{r_f,m,k}\right)\log_2(q),\\
\mathsf{size_\sigma} &= \frac{l}{8}\left(2k(1+m) + b_{r_a,m,k} + b_{r_z,m,k}\right)\log_2(q).
\end{aligned}$$

Table 3. Parameters for MURAVE, with the following irreducible polynomials $P(X)$: $P_{67} = X^{67} + X^5 + X^2 + X + 1$, $P_{83} = X^{83} + X^7 + X^4 + X^2 + 1$ and $P_{131} = X^{131} + X^8 + X^3 + X^2 + 1$.

Schemes	$(l, q, m, k, r_\lambda, r_u, r_a, r_b)$	$P(X)$	$\mathsf{size_{pk}}$	$\mathsf{size_{sk}}$	$\mathsf{size_\sigma}$
MURAVE-1	(4, 2, 79, 67, 5, 3, 23, 5)	P_{67}	5.33 KB	1.24 KB	9.69 KB
MURAVE-2	(4, 2, 92, 83, 6, 3, 26, 7)	P_{83}	8.08 KB	1.62 KB	14.50 KB
MURAVE-3	(5, 2, 149, 83, 5, 3, 33, 15)	P_{83}	15.49 KB	2.47 KB	28.08 KB
MURAVE-4	(3, 2, 139, 131, 5, 4, 57, 15)	P_{131}	13.69 KB	2.13 KB	26.35 KB

For the security level of the schemes, we consider all the attacks on the MURAVE signature scheme as stated in Sect. 5. In particular, we will evaluate the complexities of the following in Table 4:

(1) $q^{-k}\mathcal{O}_{\mathsf{RSD}}(q, m, 2k, k, 2r_u r_\lambda)$ since $2r_u r_\lambda = r_w(r_u + r_v)$.
(2) $\mathcal{O}_{\mathsf{RSD}}(q, m, 2k, k, r_u(1 + r_u))$ since $r_u(1 + r_u) = r_e(1 + r_f)$.
(3) $q^{-k}\mathcal{O}_{\mathsf{RSD}}(q, m, 2k, k, 2r_\lambda)$ since $2r_\lambda = r_w + r_\lambda$.
(4) $q^{-k}\mathcal{O}_{\mathsf{RSD}}(q, m, 2k, k, r_u(r_\lambda + 1))$ since $r_u(r_\lambda + 1) = r_\lambda r_u + r_e = r_e + r_w r_u$.
(5) $\mathsf{RSBD}(q, m, r_\lambda r_u, r_u)$ since $r_f = r_u$.

Note that the classical security level (denoted as "$\mathsf{Sec}_{\mathsf{CL}}$") of the schemes are evaluated using formulas for the complexities of the combinatorial attacks (denoted as "CmB") and algebraic attacks (denoted as "Alg") in Table 1 and 2 respectively. Bernstein [12] showed that the exponential term in the decoding complexity should be square rooted using Grover's algorithm with quantum computer. Therefore, for the post-quantum security level (denoted as "$\mathsf{Sec}_{\mathsf{PQ}}$") of the schemes, the solving complexities for the combinatorial attacks ("CmB" in Table 5) should be evaluated by taking square root of the exponential term of the formulas (in Table 1).

Table 4. Classical security level for MURAVE and complexities of attacks

Schemes	(1)		(2)		(3)		(4)		(5)	$\mathsf{Sec}_{\mathsf{PQ}}$
	CmB	Alg	CmB	Alg	CmB	Alg	CmB	Alg		
MURAVE-3	1094	716	439	321	292	203	613	408	152	128
MURAVE-4	1626	886	531	332	448	249	889	488	188	128

Table 5. Post-quantum security level for MURAVE and complexities of attacks

Schemes	(1)		(2)		(3)		(4)		(5)	$\mathsf{Sec}_{\mathsf{PQ}}$
	CmB	Alg	CmB	Alg	CmB	Alg	CmB	Alg		
MURAVE-3	1094	716	439	321	292	203	613	408	152	128
MURAVE-4	1626	886	531	332	448	249	889	488	188	128

Comparisons of Key Sizes and Signature Sizes. We compare the performance of our signature scheme, MURAVE with other rank metric signature schemes in terms of key sizes and signature sizes (Table 6).

Table 6. Comparisons of key sizes and signature sizes.

Schemes	$\mathsf{size}_{\mathsf{pk}}$	$\mathsf{size}_{\mathsf{sk}}$	size_{σ}	Sec	Remark
Durandal-I	15.25 KB	2.565 KB	4.06 KB	128	Schnorr (with FS Aborts)
cRVDC	0.152 KB	0.151 KB	22.48 KB	125	Fiat-Shamir Transformation
CVE	7.638 KB	0.210 KB	436.6 KB	80	Fiat-Shamir Transformation
MURAVE-1	5.33 KB	1.24 KB	9.69 KB	128	Schnorr (w/o FS Aborts)

Overall, our signature scheme outperforms Durandal in terms of $\mathsf{size}_{\mathsf{pk}}$ and $\mathsf{size}_{\mathsf{sk}}$, whilst our signature size is about twice the signature size of Durandal. Furthermore, our signature has smaller signature size than cRVDC and CVE.

Comparisons of Implementation Timings. Furthermore, we compare the implementation efficiency of MURAVE and Durandal. To get a good comparison between MURAVE and Durandal, our implementation is modified based on the implementation codes for Durandal (available at [1]), which is implemented in C++ and based on NTL and GMP. We implement both MURAVE and Durandal on an Intel(R) Core(TM) i7-6700 CPU at 3.40 GHz processor with 16GB of memory. The computation times for both MURAVE and Durandal signature scheme are calculated based on the average timing for 1000 instances. Note that the offline phase in $\mathsf{Sign}(\boldsymbol{m}, \mathsf{pk}, \mathsf{sk})$ refers to all the steps before Step 2, while the online phase in $\mathsf{Sign}(\boldsymbol{m}, \mathsf{pk}, \mathsf{sk})$ starts from Step 2 (the computation of $\boldsymbol{c}$) (Table 7).

Table 7. Comparisons of implementation results.

Schemes	Key.Gen	Offline Sign Phase	Online Sign Phase	Verify
Durandal-I	4.7 ms	732.7 ms	8.4 ms	7.6 ms
MURAVE-1	4.6 ms	14.5 ms	3.4 ms	6.9 ms

Our implementation results show that MURAVE is much more efficient than Durandal, especially in the $\mathsf{Sign}(\boldsymbol{m}, \mathsf{pk}, \mathsf{sk})$ phase. Our implementations confirm the efficiency, practicality and advantages of our signature scheme, as we use only random vectors $\boldsymbol{\lambda}_i$ to mask the secret key which involves only polynomial additions and multiplications. Hence, our signature generation is much faster than Durandal, as the latter requires costly computation in solving a linear system. Furthermore, each steps in our signature generation is deterministic, i.e., we are not required to apply the Fiat-Shamir with Aborts strategy. As such, our signature generation (within 17.9 ms) can be much more efficient than signature generation in Durandal (within 741.1 ms).

7 Conclusion

We have proposed a new rank metric signature scheme called MURAVE constructed via the Schnorr approach. Our signature scheme is efficient as it involves only polynomial additions and multiplications. Our implementations confirm the efficiency of our signature scheme as it requires 17.9 ms on average to produce a valid signature. Furthermore, we have introduced some new problems in rank metric code-based cryptography, namely the Rank Support Basis Decomposition and the Decisional Advanced Rank Support Basis Decomposition (DRSBD^+) problem. We have given a proof in the EUF-CMA security model, reducing the security of the scheme to the Rank Syndrome Decoding (I-RSD) problem for ideal codes, the Ideal LRPC Codes Indistinguishability problem and DRSBD^+ problem. We have shown that our scheme is secure against known attacks on rank metric signature schemes. Finally, we have made comparison with other existing rank metric signature schemes that are still secure up-to-date. Our scheme is efficient in terms of key sizes (of order 5.33 KB) and of signature sizes (of 9.69 KB) for 128-bit classical security level.

Acknowledgement. We are grateful to the anonymous reviewers for their careful reading of our manuscript and their many insightful comments and suggestions which have greatly improved this manuscript.

Appendix A Rank Support Recovery Algorithm

Let $f = (f_1, \ldots, f_d) \in E_{m,d,d}$, $e = (e_1, \ldots, e_r) \in E_{m,r,r}$ and $\boldsymbol{s} = (s_1, \ldots, s_n) \in \mathbb{F}_{q^m}^n$ such that $S := \langle s_1, \ldots, s_n \rangle = \langle f_1 e_1, \ldots, f_d e_r \rangle$. Given $\boldsymbol{f}$, $\boldsymbol{s}$ and r as input, the Rank Support Recovery Algorithm will output a vector space E which satisfies $E = \langle e_1, \ldots, e_r \rangle$. Denote $S_i := f_i^{-1}.S$ and $S_{i,j} := S_i \cap S_j$.

Algorithm 1: Rank Support Recovery (RSR) Algorithm

Data: $F = \langle f_1, \ldots, f_d \rangle$, $\boldsymbol{s} = (s_1, \ldots, s_n) \in \mathbb{F}_{q^m}^n$, $r = \dim(E)$
Result: A candidate for the vector space E

Compute $S = \langle s_1, \ldots, s_n \rangle$
Precompute every S_i for $i = 1$ to d
Precompute every $S_{i,i+1}$ for $i = 1$ to $d - 1$
for $i \leftarrow 1$ **to** $d - 2$ **do**
  $tmp \leftarrow S + F.(S_{i,i+1} + S_{i+1,i+2} + S_{i,i+2})$
  if $\dim(tmp) \leq rd$ **then**
    $S \leftarrow tmp$
$E \leftarrow f_1^{-1}.S \cap \ldots \cap f_d^{-1}.S$
return E

References

1. Aragon, N.: Durandal Implementation, Github, 10 May 2019. https://github.com/nicolas-aragon/Durandal
2. Aragon, N., et al.: Cryptanalysis of a rank-based signature with short public keys. Des. Codes Crypt. **88**, 643–653 (2020)
3. Aragon, N., Blazy, O., Gaborit, P., Hauteville, A., Zémor, G.: Durandal: a rank metric based signature scheme. In: Ishai, Y., Rijmen, V. (eds.) EUROCRYPT 2019. LNCS, vol. 11478, pp. 728–758. Springer, Cham (2019). https://doi.org/10.1007/978-3-030-17659-4_25
4. Aragon, N., Gaborit, P., Hauteville, A., Ruatta, O., Zémor, G.: Low rank parity check codes: new decoding algorithms and applications to cryptography. IEEE Trans. Inf. Theory **65**(12), 7697–7717 (2019)
5. Aragon, N., Gaborit, P., Hauteville, A., Tillich, J.-P.: A new algorithm for solving the rank syndrome decoding problem. In: IEEE International Symposium on Information Theory (ISIT 2018), pp. 2421–2425 (2018)

6. Aragon, N., Ruatta, O., Gaborit, P., Zémor, G., Hauteville, A.: RankSign - a signature proposal for the NIST's call, Specification vision 1.0, 30 November 2017. https://csrc.nist.gov/Projects/Post-Quantum-Cryptography/Round-1-Submissions
7. Bardet, M., et al.: An algebraic attack on rank metric code-based cryptosystems. In: Canteaut, A., Ishai, Y. (eds.) EUROCRYPT 2020. LNCS, vol. 12107, pp. 64–93. Springer, Cham (2020). https://doi.org/10.1007/978-3-030-45727-3_3
8. Bardet, M., et al.: Algebraic attacks for solving the Rank Decoding and MinRank problems without Gröbner basis. CoRR abs/2002.08322 (2020)
9. Bellini, E., Caullery, F., Hasikos, A., Manzano, M., Mateu, V.: Code-based signature schemes from identification protocols in the rank metric. In: Camenisch, J., Papadimitratos, P. (eds.) CANS 2018. LNCS, vol. 11124, pp. 277–298. Springer, Cham (2018). https://doi.org/10.1007/978-3-030-00434-7_14
10. Bellini, E., Caullery, F., Gaborit, P., Manzano, M., Mateu, V.: Improved veron identification and signature schemes in the rank metric. In: IEEE International Symposium on Information Theory (ISIT 2019), pp. 1872–1876 (2019)
11. Berlekamp, E.E., McEliece, R., Tilborg, H.V.: On the inherent intractability of certain coding problems. IEEE Trans. Inf. Theory **24**(3), 384–386 (1978)
12. Bernstein, D.J.: Grover vs. McEliece. In: Sendrier, N. (ed.) PQCrypto 2010. LNCS, vol. 6061, pp. 73–80. Springer, Heidelberg (2010). https://doi.org/10.1007/978-3-642-12929-2_6
13. Bernstein, D.J., Hülsing, A., Lange, T., Panny, L.: Comments on RaCoSS, a submission to NIST's PQC competition, 23 December 2017. https://helaas.org/racoss/
14. Chabaud, F., Stern, J.: The cryptographic security of the syndrome decoding problem for rank distance codes. In: Kim, K., Matsumoto, T. (eds.) ASIACRYPT 1996. LNCS, vol. 1163, pp. 368–381. Springer, Heidelberg (1996). https://doi.org/10.1007/BFb0034862
15. Debris-Alazard, T., Sendrier, N., Tillich, J.-P.: Wave: a new family of trapdoor one-way preimage sampleable functions based on codes. In: Galbraith, S.D., Moriai, S. (eds.) ASIACRYPT 2019. LNCS, vol. 11921, pp. 21–51. Springer, Cham (2019). https://doi.org/10.1007/978-3-030-34578-5_2
16. Debris-Alazard, T., Tillich, J.-P.: Two attacks on rank metric code-based schemes: RankSign and an IBE scheme. In: Peyrin, T., Galbraith, S. (eds.) ASIACRYPT 2018. LNCS, vol. 11272, pp. 62–92. Springer, Cham (2018). https://doi.org/10.1007/978-3-030-03326-2_3
17. Faugère, J.-C., Levy-dit-Vehel, F., Perret, L.: Cryptanalysis of MinRank. In: Wagner, D. (ed.) CRYPTO 2008. LNCS, vol. 5157, pp. 280–296. Springer, Heidelberg (2008). https://doi.org/10.1007/978-3-540-85174-5_16
18. Gaborit, P., Ruatta, O., Schrek, J.: On the complexity of the rank syndrome decoding problem. IEEE Trans. Inf. Theory **62**(2), 1006–1019 (2016)
19. Gaborit, P., Zémor, G.: On the hardness of the decoding and the minimum distance problems for rank codes. IEEE Trans. Inf. Theory **62**(12), 7245–7252 (2016)
20. Goldwasser, S., Micali, S., Rivest, R.: A digital signature scheme secure against adaptive chosen-message attacks. SIAM J. Comput. **17**(2), 281–308 (1988)
21. Goubin, L., Courtois, N.T.: Cryptanalysis of the TTM cryptosystem. In: Okamoto, T. (ed.) ASIACRYPT 2000. LNCS, vol. 1976, pp. 44–57. Springer, Heidelberg (2000). https://doi.org/10.1007/3-540-44448-3_4
22. Hastad, J., Impagliazzo, R., Levin, L., Luby, M.: A pseudorandom generator from any one-way function. SIAM J. Comput. **28**(4), 1364–1396 (1999)

23. Horlemann-Trautmann, A., Marshall, K., Rosenthal, J.: Extension of overbeck's attack for gabidulin based cryptosystems. Des. Codes Crypt. **86**(2), 319–340 (2018)
24. Lau, T.S.C., Tan, C.H., Prabowo, T.F.: Key recovery attacks on some rank metric code-based signatures. In: Albrecht, M. (ed.) IMACC 2019. LNCS, vol. 11929, pp. 215–235. Springer, Cham (2019). https://doi.org/10.1007/978-3-030-35199-1_11
25. Lee, W., Kim, Y.S., Lee, Y.W., No, J.S.: Post quantum signature scheme based on modified Reed-Muller code (pqsigRM), Specification vision 1.0, 30 November 2017. https://csrc.nist.gov/Projects/Post-Quantum-Cryptography/Round-1-Submissions
26. Levy-dit-Vehel, F., Perret, L.: Algebraic decoding of rank metric codes. In: Yet Another Conference on Cryptography (YACC 2006), pp. 142–152 (2006)
27. Lyubashevsky, V.: Fiat-Shamir with aborts: applications to lattice and factoring-based signatures. In: Matsui, M. (ed.) ASIACRYPT 2009. LNCS, vol. 5912, pp. 598–616. Springer, Heidelberg (2009). https://doi.org/10.1007/978-3-642-10366-7_35
28. Ourivski, A.V., Johansson, T.: New technique for decoding codes in the rank metric and its cryptography applications. Prob. Inf. Transm. **38**(3), 237–246 (2002)
29. Roy, P.S., Xu, R., Fukushima, K., Kiyomoto, S., Morozov, K., Takagi, T.: Random code-based signature scheme, Specification vision 1.0, 29 November 2017. https://csrc.nist.gov/Projects/Post-Quantum-Cryptography/Round-1-Submissions
30. Schnorr, C.P.: Efficient identification and signatures for smart cards. In: Brassard, G. (ed.) CRYPTO 1989. LNCS, vol. 435, pp. 239–252. Springer, New York (1990). https://doi.org/10.1007/0-387-34805-0_22
31. Song, Y., Huang, X., Mu, Y., Wu, W.: A New Code-based Signature Scheme with Shorter Public Key. Cryptology ePrint Archive: Report 2019/053. https://eprint.iacr.org/eprint-bin/getfile.pl?entry=2019/053&version=20190125:204017&file=053.pdf
32. Tan, C.H., Prabowo, T.F., Lau, T.S.C.: Rank metric code-based signature. In: IEEE International Symposium on Information Theory and Its Application (ISITA 2018), pp. 70–74 (2018)

Optimized and Secure Implementation of ROLLO-I

Lina Mortajine[1,3(✉)], Othman Benchaalal[1], Pierre-Louis Cayrel[2], Nadia El Mrabet[3], and Jérôme Lablanche[1]

[1] Wisekey, Arteparc de Bachasson, Bâtiment A, 13590 Meyreuil, France
{lmortajine,obenchaalal,jlablanche}@wisekey.com
[2] Laboratoire Hubert Curien, UMR CNRS 5516, Bâtiment F 18 rue du Benoît Lauras, 42000 Saint-Etienne, France
pierre.louis.cayrel@univ-st-etienne.fr
[3] Mines Saint-Etienne, CEA-Tech, Centre CMP, Departement SAS, 13541 Gardanne, France
nadia.el-mrabet@emse.fr

Abstract. This paper presents our contribution regarding two implementations of the ROLLO-I algorithm, a code-based candidate for the NIST PQC project. The first part focuses on the implementations, and the second part analyzes a side-channel attack and the associated countermeasures. The first implementation uses existing hardware with a crypto co-processor to speed-up operations in $\mathbb{F}_{2^m}$. The second one is a full software implementation (not using the crypto co-processor), running on the same hardware. Finally, the side-channel attack allows us to recover the secret key with only 79 ciphertexts for ROLLO-I-128. We propose countermeasures in order to protect future implementations.

Keywords: Post-quantum cryptography · Side-channel attacks · ROLLO-I cryptosystem

Introduction

Today, 26 candidates are still under study for the standardization campaign launched by the National Institute of Standards and Technology (NIST) in 2016. Among the candidates that were submitted are 8 signature schemes based on lattices and multivariate. Also submitted were 17 public-key encryption schemes, key-encapsulation mechanisms (KEMs), that base their security on codes, lattices, or isogenies. In addition, one more signature scheme based on a zero-knowledge proof system has also been submitted.

In this paper, we focus our analysis on the submissions based on codes. The first cryptosystems based on codes (e.g. McEliece cryptostem [17]) uses keys far too large to be usable by the industry. The development of new cryptosystems based on different codes as well as the introduction of codes embedded with the rank metric has resulted in a considerable reduction of key sizes and thus

M. Baldi et al. (Eds.): CBCrypto 2020, LNCS 12087, pp. 117–137, 2020.
https://doi.org/10.1007/978-3-030-54074-6_7

reaches key sizes comparable to those used in lattice-based cryptography. Despite the evolution of research in this field, some post-quantum cryptosystems submitted to the NIST PQC project require a large number of resources. Notably regarding the memory which becomes binding when we have to implement the algorithms into constrained environments such that microcontrollers. It is then hardly conceivable to imagine that these cryptosystems may replace the ones used nowadays on chips. In that purpose, we decided to study the real cost of a code-based cryptosystem implementation. This study is essential to prepare the transition to post-quantum cryptography. For this study, we decided to perform two implementations on microcontroller, the first one using only software and the second one using the crypto co-processor featuring in the microcontroller.

One of the main criteria for the selection of the cryptosystem has been the RAM available on the microcontroller to run cryptographic protocols. We first decided to compare the size of elements manipulated in submitted code-based cryptosystems. The respective sizes are reported in Table 1. Three other code-based cryptosystems in round 2; Classic McEliece, LEDAcrypt, and NTS-KEM use much larger keys and, thus were not taken into account in our study and not listed in Table 1.

Table 1. Size of elements in bytes for code-based cryptosystems (security level 5)

Parameter	Algorithm							
	BIKE [14]			HQC [15]	RQC [16]	ROLLO		
Scheme number	I	II	III			**I**	II	III
Public key	8,188	4,094	9,033	14,754	3,510	**947**	2,493	2,196
Secret key	548	548	532	532	3,510	**1,894**	4,986	2,196
Ciphertext	8,188	4,094	9,033	14,818	3,574	**947**	2,621	2,196

The selection of a microcontroller with only 4 kB of RAM that can be found on the market led us to choose ROLLO-I submission. Indeed, as seen in Table 1, the total size of its parameters is the smallest one. Thus, we will suppose that its algorithm needs the smallest amount of RAM. Since operations on ROLLO-II and ROLLO-III are similar, they should be integrated quickly.

Embedded implementations can lead to vulnerabilities that a side-channel attacker can exploit. He gathers information about private data by exploiting physical measurements. Some side-channel attacks have already been performed on code-based cryptosystems [1,2]. Then, to provide a first secure implementation of ROLLO-I, we propose the countermeasures against a side-channel attack that we introduce.

Our Contribution. We present two practical implementations of ROLLO-I in a microcontroller in which 4 kB of RAM is dedicated to cryptographic data. The first one consisting in full software implementation and the second one uses the crypto co-processor featuring in the microcontroller.

We finally give a first study on the security of ROLLO-I against side-channel attacks and implement countermeasures against the attack that we have found.

Organization of this Paper. We start with some preliminary definitions and present ROLLO-I cryptosystem in Sect. 1, then we present in Sect. 2 the memory-optimized implementations and in Sect. 3, we finally demonstrate a first side-channel attack on ROLLO-I and present associated countermeasures.

1 Background

In this section, we give some definitions to explain the Low-Rank Parity Check (LRPC) codes which have been first introduced in [3]. For more details, the reader is referred to [4]. For fixed positive integers m and n, we denote by:

q	a power of a prime number p
$\mathbb{F}_q$	the finite field of q elements
$\mathbb{F}_{q^m}$	the vector space that is isomorphic to $\mathbb{F}_q[x]/(P_m)$, with P_m an irreducible polynomial of degree m over $\mathbb{F}_q$
$\mathbb{F}_{q^m}^n$	a vector space isomorphic to $\mathbb{F}_{q^m}[X]/(P_n)$, with P_n an irreducible polynomial of degree n over $\mathbb{F}_q$
$\mathbf{v}$	an element of $\mathbb{F}_{q^m}^n$
$M(\mathbf{v})$	the matrix $(v_{i,j})_{\substack{1\leq i\leq n\\ 1\leq j\leq m}}$ corresponding to the element $\mathbf{v}$.

Let k be an integer. A linear code $\mathcal{C}$ over $\mathbb{F}_{q^m}$ of length n and dimension k is a subspace of $\mathbb{F}_{q^m}^n$. It is denoted by $[n,k]_{q^m}$, and can be represented by a generator matrix $\mathbf{G} \in \mathbb{F}_{q^m}^{k\times n}$ such that

$$\mathcal{C} := \{\mathbf{x}.\mathbf{G}, \mathbf{x} \in \mathbb{F}_{q^m}^k\}.$$

The code $\mathcal{C}$ can also be given by its parity-check matrix $\mathbf{H} \in \mathbb{F}_{q^m}^{(n-k)\times n}$ such that

$$\mathcal{C} := \{\mathbf{x} \in \mathbb{F}_{q^m}^n, \mathbf{H}.\mathbf{x}^T = 0\}.$$

The vector $\mathbf{s_x} = \mathbf{H}.\mathbf{x}^T$ is called the syndrome of $\mathbf{x}$.

ROLLO cryptosystem is based on codes in rank metric over $\mathbb{F}_{q^m}^n$. In rank metric, the distance between two words $\mathbf{x} = (x_1, \cdots, x_n)$ and $\mathbf{y} = (y_1, \cdots, y_n)$ in $\mathbb{F}_{q^m}^n$ is defined by

$$d(\mathbf{x}, \mathbf{y}) := \|\mathbf{x} - \mathbf{y}\| = \|\mathbf{v}\| = \text{Rank } M(\mathbf{v}),$$

where $\|\mathbf{v}\|$ is the rank weight of the word $\mathbf{v} = \mathbf{x} - \mathbf{y}$.

The rank of a word $\mathbf{x} = (x_1, \cdots, x_n)$ can also be seen as the dimension of its support $\text{Supp}(\mathbf{x}) \subset \mathbb{F}_{q^m}$ spanned by the basis of $\mathbf{x}$. Namely, the support of $\mathbf{x}$ is given by

$$\text{Supp}(\mathbf{x}) = \langle x_1, \cdots, x_n\rangle_{\mathbb{F}_q}.$$

The authors of [4] introduced the family of ideal codes that allows them to reduce the size of the code's representation. The associated generator matrix is based on ideal matrices.

Given a polynomial $P \in \mathbb{F}_q[X]$ of degree n and a vector $\mathbf{v} \in \mathbb{F}_{q^m}^n$, an ideal matrix generated by $\mathbf{v}$ is an $n \times n$ square matrix defined by

$$\mathcal{IM}(\mathbf{v}) = \begin{pmatrix} \mathbf{v} \\ X\mathbf{v} \bmod P \\ \vdots \\ X^{n-1}\mathbf{v} \bmod P \end{pmatrix}.$$

An $[ns, nt]_{q^m}$-code $\mathcal{C}$, generated by the vectors $(\mathbf{g}_{i,j})_{\substack{i\in[1,\cdots,s-t] \\ j\in[1,\cdots,t]}} \in \mathbb{F}_{q^m}^n$, is an ideal code if a generator matrix in systematic form is of the form

$$\mathbf{G} = \begin{pmatrix} & \mathcal{IM}(\mathbf{g_{1,1}}) & \cdots & \mathcal{IM}(\mathbf{g_{1,s-t}}) \\ \mathbf{I}_{nt} & \vdots & \ddots & \vdots \\ & \mathcal{IM}(\mathbf{g_{t,1}}) & \cdots & \mathcal{IM}(\mathbf{g_{t,s-t}}) \end{pmatrix}.$$

In [4], the authors restrain the definition of ideal LRPC (Low-Rank Parity Check) codes to (2, 1)-ideal LRPC codes that they used in ROLLO cryptosystems.

Let F be a $\mathbb{F}_q$-subspace of $\mathbb{F}_{q^m}$ such that $dim(F) = d$. Let $(\mathbf{h}_1, \mathbf{h}_2)$ be two vectors of $\mathbb{F}_{q^m}^n$, such that $\mathrm{Supp}(\mathbf{h}_1, \mathbf{h}_2) = F$, and $P \in \mathbb{F}_q[X]$ be a polynomial of degree n. A $[2n, n]_{q^m}$-code $\mathcal{C}$ is an ideal LRPC code if it has a parity-check matrix of the form

$$\mathbf{H} = \begin{pmatrix} \mathcal{IM}(\mathbf{h}_1)^T & \mathcal{IM}(\mathbf{h}_2)^T \end{pmatrix}.$$

Hereafter, we will focus on ROLLO-I submission, which has smaller parameters than ROLLO-II and ROLLO-III (see Table 1).

ROLLO-I Scheme

The submission of ROLLO-I is a Key Encapsulation Mechanism (KEM) composed of three probabilistic algorithms: the Key generation (Keygen), Encapsulation (Encap), and Decapsulation (Decap) are detailed in Table 3. During the decapsulation process, the syndrome of the received ciphertext c is computed, then the Rank Support Recovery (RSR) algorithm is performed to recover the error's support. The latter is explained in [4].

The fixed parameter sets given in Table 2 allow to achieve respectively 128, 192, and 256-bit security level according to NIST's security strength categories 1, 3, and 5 [5]. As described in Sect. 1, the parameters n and m correspond respectively to the degrees of irreducible polynomials P_n and P_m implied in the fields $\mathbb{F}_q[x]/(P_m)$ and $F_{q^m}[X]/(P_n)$. We note that for the three security levels, $q = 2$. The parameters d and r correspond respectively to the private key and error's ranks.

Table 2. ROLLO-I parameters for each security level

Algo.	Param.				
	d	r	P_n	P_m	Security level (bits)
ROLLO-I-128	6	5	$X^{47}+X^5+1$	$x^{79}+x^9+1$	128
ROLLO-I-192	7	6	$X^{53}+X^6+X^2+X+1$	$x^{89}+x^{38}+1$	192
ROLLO-I-256	8	7	$X^{67}+X^5+X^2+X+1$	$x^{113}+x^9+1$	256

Table 3. ROLLO-I KEM protocol

Alice		Bob
KeyGen Generate a support F of rank d Generate the private key $\mathbf{sk} = (\mathbf{x}, \mathbf{y})$ from the support F Compute the public key $\mathbf{h} = \mathbf{x}^{-1} \cdot \mathbf{y} \mod P_n$	$\xrightarrow{\mathbf{h}}$	**Encapsulation** Generate a support E of rank r Pick randomly two elements $(\mathbf{e_1}, \mathbf{e_2})$ from the support E Compute the ciphertext $\mathbf{c} = \mathbf{e_2} + \mathbf{e_1} \cdot \mathbf{h} \mod P_n$ Derive the shared secret $K = \text{Hash}(E)$
Decapsulation Compute the syndrome $\mathbf{s} = \mathbf{x} \cdot \mathbf{c} \mod P_n = x.\mathbf{e_2} + y.\mathbf{e_1} \mod P_n$ Recover the error's support $E = \text{RSR}(F, \mathbf{s}, r)$ Compute the shared secret $K = \text{Hash}(E)$	$\xleftarrow{\mathbf{c}}$	

2 ROLLO-I Implementations

In this section we detail the algorithms in the rings $\mathbb{F}_2[x]/(P_m)$ and $\mathbb{F}_{2^m}[X]/(P_n)$ required in ROLLO-I cryptosystem. The implementations are performed on 32-bit architecture systems.

2.1 Operations in $\mathbb{F}_2[x]/(P_m)$

The addition in $\mathbb{F}_2[x]/(P_m)$ consists in xoring 32-bit words. Thus, the three main operations to implement are the multiplication, the modular reduction, and the inversion. For the inversion in $\mathbb{F}_2[x]/(P_m)$, we use the extended Euclidean algorithm for binary polynomials as given in [6, Algo. 2.48].

2.1.1 Multiplication

Regarding the multiplication between two polynomials $a, b \in \mathbb{F}_2[x]/(P_m)$, we use the left-to-right comb method with windows of width $w = 4$ as described in [6, Algo. 2.36]. For any polynomial $a \in \mathbb{F}_2[x]/(P_m)$, we associate the vector $A = (A_0, \ldots, A_{\lceil m/32 \rceil - 1})$ where A_j is the jth 32-bit word and we note $A_{j,i}$ the ith block of four coefficients in A_j. First we pre-compute the product $u(x) \times b(x)$ for all polynomials u of degree less than 4 (16 elements are stored in a table T). Let $\hat{u}$ denote the binary representation of the coefficients of polynomial $u(x)$ (i.e $u(x) = 0 \leftrightarrow \hat{u} = 0, u(x) = 1 \leftrightarrow \hat{u} = 1, u(x) = x \leftrightarrow \hat{u} = 2, \cdots, u(x) = x^3 + x^2 + x + 1 \leftrightarrow \hat{u} = 15$). Thus, we have $T_{\hat{u}} = b(x) \times u(x)$.

Then, for $0 \leq j < \lceil m/32 \rceil$, we add to the result $\mathbf{R_j} = (R_j, \ldots, R_n)$, the element $T_{\hat{u}}$, where $\hat{u}$ is the integer associated to $A_{j,i}$, for each i. If i is non zero, we multiply the polynomial $\mathbf{R}$ by x^4, which is equivalent to a shift of 32-bit words.

Algorithm 1: Polynomial multiplication using the left-to-right method with a width window $w = 4$

Input: Two polynomials $a, b \in \mathbb{F}_2[x]/(P_m)$
Output: $r(x) = a(x) \times b(x)$

For all polynomials $u(x)$ of degree at most $w - 1$, compute $T_{\hat{u}} = b(x) \times u(x)$
$\mathbf{R} \leftarrow 0$
for i *from* 7 *downto* 0 **do**
  for j *from 0 to* $\lceil m/32 \rceil - 1$ **do**
    Let $\hat{u} = u_3u_2u_1u_0$ where u_k is the bit $wi + k$ of A_j.
    $\mathbf{R_j} \leftarrow \mathbf{R_j} \oplus T_{\hat{u}}$
  if $i \neq 0$ **then**
    $R(x) \leftarrow R(x) \times x^4$
return R

2.1.2 Modular Reduction

Several modular reductions with parse polynomials are performed in ROLLO-I cryptosystem. We decide to use the same technique explained in [6, Sect. 2.3.5].

Let us take the example of ROLLO-I-128 and consider an element $\mathbf{c} = (c_i)_{0 \leq i \leq 156}$ obtained after a multiplication of two elements in $\mathbb{F}_2[x]/(P_{79})$. The modular reduction is performed on each C_i 32-bit word composing $\mathbf{c}$ with $0 \leq i \leq 4$, as in Algorithm 2.

Allow us detail the method for the reduction modulo $P_m(x) = x^{79} + x^9 + 1$ of the 4th word of $\mathbf{c}$ which corresponds to the polynomial $c_{96}x^{96} + c_{97}x^{97} + \cdots + c_{127}x^{127}$.

We have:

$$x^{96} \equiv x^{17} + x^{26} \mod P_m$$
$$\vdots$$
$$x^{127} \equiv x^{48} + x^{57} \mod P_m$$

Given those congruences, the reduction of C_3 is operated by adding two times C_3 to $\mathbf{c}$ as shown in Fig. 1.

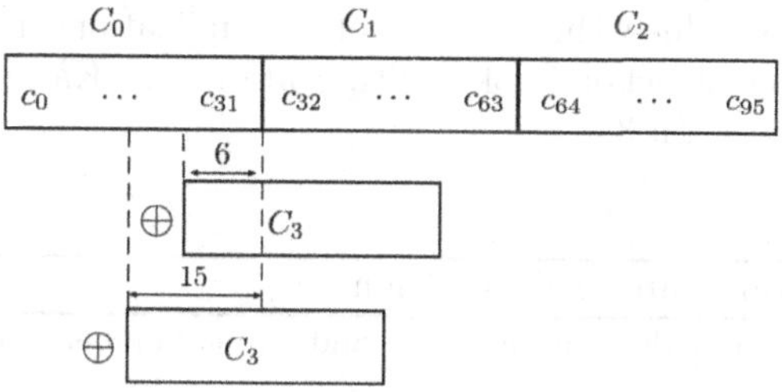

Fig. 1. Reduction of the 32-bit word C_3 modulo $P_m(x) = x^{79} + x^9 + 1$

Algorithm 2: Reduction modulo $P_m(x) = x^{79} + x^9 + 1$

Input: polynomial $c(x)$ of degree at most 156
Output: $c(x) \mod P_m(x)$

// $C_i = (c_{31+32\times i} \cdots c_{32\times i})$
$C_2 \leftarrow C_2 \oplus (C_4 \gg 6) \oplus (C_4 \gg 15)$
$C_1 \leftarrow C_1 \oplus (C_4 \ll 17) \oplus (C_4 \ll 26) \oplus (C_3 \gg 6) \oplus (C_3 \gg 15)$
$C_0 \leftarrow C_0 \oplus (C_3 \ll 17) \oplus (C_3 \ll 26)$
$T \leftarrow C_2$ & 0xFFFF8000
$C_0 \leftarrow C_0 \oplus (T \gg 15) \oplus (T \gg 6)$
$C_2 \leftarrow C_2$ & 0x7FFF
$C_3, C_4 \leftarrow 0$
return C

2.2 Operations and Memory Costs Issues in $\mathbb{F}_{2^m}[X]/(P_n)$

In this section, m_b represents the length in bytes of coefficients in $\mathbb{F}_{2^m}$.

2.2.1 Multiplication

The multiplication in $\mathbb{F}_{2^m}[X]/(P_n)$ is one of the most used operations of this cryptosystem: it is involved in the computation of the public key, the ciphertext and the syndrome.

For example, let $P(X) = p_0 + p_1 X$ and $Q(X) = q_0 + q_1 X$ be two polynomials of degree 1 in a given polynomial ring. The result of the product is

$$P(X) \times Q(X) = p_0 q_0 + (p_0 q_1 + p_1 q_0)X + p_1 q_1 X^2.$$

Naively, we have four multiplications and one addition over the coefficients. Thus, the schoolbook multiplication [7] requires n^2 multiplications in $\mathbb{F}_{2^m}$. The Karatsuba algorithm uses the following equation

$$(p_0 q_1 + p_1 q_0) = (p_0 + p_1)(q_0 + q_1) - p_0 q_0 - p_1 q_1,$$

and $P(X) \times Q(X)$ requires only three multiplications and four additions over the coefficients. To reduce the number of multiplications in $\mathbb{F}_{2^m}$, we implement a combination of Schoolbook multiplication and Karatsuba method [8], as described in Algorithm 3.

Algorithm 3: Karatsuba multiplication

Input: two polynomials $\mathbf{f}$ and $\mathbf{g} \in \mathbb{F}_{2^m}^n$ and N the number of coefficients of $\mathbf{f}$ and $\mathbf{g}$

Output: $\mathbf{f} \cdot \mathbf{g}$ in $\mathbb{F}_{2^m}^n$

1 **if** N *odd* **then**

2   *result* $\leftarrow$ Schoolbook$(\mathbf{f}, \mathbf{g}, N)$

3   **return** *result*

4 $N' \leftarrow N/2$

5 Let $\mathbf{f}(x) = \mathbf{f}_0(x) + \mathbf{f}_1(x)x^{N'}$

6 Let $\mathbf{g}(x) = \mathbf{g}_0(x) + \mathbf{g}_1(x)x^{N'}$

7 $R_1 \leftarrow$ Karatsuba$(\mathbf{f}_0, \mathbf{g}_0, N')$ // Compute recursively $\mathbf{f}_0\mathbf{g}_0$

8 $R_2 \leftarrow$ Karatsuba$(\mathbf{f}_1, \mathbf{g}_1, N')$ // Compute recursively $\mathbf{f}_1\mathbf{g}_1$

9 $R_3 \leftarrow \mathbf{f}_0 + \mathbf{f}_1$

10 $R_4 \leftarrow \mathbf{g}_0 + \mathbf{g}_1$

11 $R_5 \leftarrow$ Karatsuba(R_3, R_4, N') // Compute recursively $R_3 R_4$

12 $R_6 \leftarrow R_5 - R_1 - R_2$

13 **return** $R_1 + R_6 x^{N'} + R_2 x^{2N}$

In line 4 (Algorithm 3), we divide the polynomial's length N by 2. Consequently, we need to add a padding to the input polynomials with zero coefficients to obtain N even. In Fig. 2, we observe that the cycles' number is not strictly increasing due to the division by 2.

Depending on the memory available for a multiplication in $\mathbb{F}_{2^m}[X]/(P_n)$, we can add more or less padding. For example, in ROLLO-I-128 with $n = 47$, we decide to add one zero coefficient which allows us to reduce considerably the number of cycles; however, in ROLLO-I-192 with $n = 53$, we have two possibilities: pad the polynomials with 3 or 11 coefficients. The second possibility

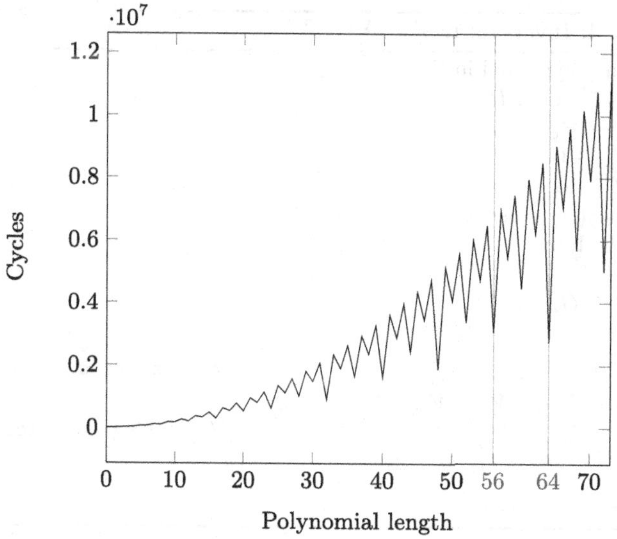

Fig. 2. Number of cycles required by Karatsuba combined with schoolbook multiplication depending on the polynomial length

is about 10% faster but requires an additional memory cost of $11 \times \lceil 89/32 \rceil \times 4 = 132$ bytes per polynomial. That is why the first choice represents a good balance between memory and execution time.

2.2.2 Inversion

For the inversion in $\mathbb{F}_{2^m}[X]/(P_n)$, we adjust extended Euclidean algorithm given in [6, Algo. 2.48] to the ring $\mathbb{F}_{2^m}[X]/(P_n)$ as described in Algorithm 4.

During the execution of the extended Euclidean algorithm, we have in memory:

- the polynomial to be inverted Q;
- a copy of Q (in order to keep it in memory);
- the dividend;
- the two Bézout coefficients;
- three buffers used to perform intermediates operations (swap between polynomials, results of multiplications).

A way to implement it is to allocate the maximum memory size for each element. As an element can be composed of n coefficients in $\mathbb{F}_{2^m}$, the computation of the inverse in $\mathbb{F}_{2^m}[X]/(P_n)$ requires $8 \times n \times m_b$ bytes. Considering the parameters of ROLLO-I-128, ROLLO-I-192 and ROLLO-I-256, the memory usage represents respectively 4,512, 5,088, and 8,576 bytes, thus exceeding the memory size available on the target microcontroller. However, during the algorithm we notice that:

Algorithm 4: Inversion in $\mathbb{F}_{2^m}[X]/(P_n)$

Input: Q a polynomial in $\mathbb{F}_{2^m}[X]/(P_n)$
Output: $Q^{-1} \mod P_n$

$U \leftarrow Q$, $V \leftarrow P_n$
$G_1 \leftarrow 1$, $G_2 \leftarrow 0$
while $U \neq 1$ **do**
  $j \longleftarrow deg(U) - deg(V)$
  if $j < 0$ **then**
    $U \leftrightarrow V$
    $G_1 \leftrightarrow G_2$
    $j \leftarrow -j$
  $lc\_V \leftarrow V_{deg(V)-1}$ `// leading coefficient of` V
  $U \leftarrow U + X^j.(lc\_V)^{-1}.V$
  $lc\_G_2 \leftarrow G_{2_{deg(G_2)-1}}$ `// leading coefficient of` G_2
  $G_1 \leftarrow G_1 + X^j.(lc\_G_2)^{-1}.G_2$
return G_1

- the degree of the polynomial Q is at most $n-1$ and the degree of the dividend is n at the beginning of the process, both decrease during the execution;
- the degrees of the two Bézout coefficients are 0 at the beginning and increase during the process.

Thus, we decide to perform a dynamic memory allocation by setting the necessary memory space for each element at each step of the inversion process. The memory usage is reduced to 2,590, 2,904 and 4,864 bytes respectively for ROLLO-I-128, ROLLO-I-192 and ROLLO-I-256.

2.2.3 Rank Support Recovery (RSR) Algorithm

The main memory issue in the RSR algorithm, given in [4, Algo. 1], is the multiple intersections between sub-spaces over $\mathbb{F}_{2^m}^n$. Considering two sub-spaces $U = \langle u_0, u_1, \cdots, u_{n-1}\rangle$ and $V = \langle v_0, v_1, \cdots, v_{n-1}\rangle$ and their associated vectors $\mathbf{u} = (u_0, u_1, \cdots, u_{n-1})$ and $\mathbf{v} = (v_0, v_1, \cdots, v_{n-1})$ in $\mathbb{F}_{2^m}^n$. The intersection $\mathcal{I}_{U,V} = U \cap V$ is computed by following Zassenhaus algorithm [9], as described below:

- Create the block matrix $\mathcal{Z}_{\mathbf{U},\mathbf{V}} = \begin{pmatrix} M(\mathbf{u}) & M(\mathbf{u}) \\ M(\mathbf{v}) & 0 \end{pmatrix}$;
- Apply the Gaussian elimination on $\mathcal{Z}_{\mathbf{U},\mathbf{V}}$ to obtain a row echelon form matrix;
- The resulting matrix has the following shape: $\begin{pmatrix} M(\mathbf{c}) & * \\ 0 & \mathcal{I}_{U,V} \\ 0 & 0 \end{pmatrix}$,

  where $\mathbf{c} \in \mathbb{F}_{2^m}^n$.

In the initial RSR algorithm, some pre-computations are performed to avoid additional operations on data.

First, we pre-compute $S_i = f_i^{-1}S$, for $1 \leq i \leq d$, where f_i are the elements of the support F and S the support of the syndrome. As each S_i is composed of $r \times d$ coefficients in $\mathbb{F}_{2^m}$, $r \times d \times d \times m_b$ bytes are needed for S_i pre-computations.

Let $S_{i,j} = S_i \cap S_j$, for $1 \leq i < j \leq d$, composed of r elements in $\mathbb{F}_{2^m}$. For the pre-computations of these intersections, we also need to consider the memory usage induced by the Zassenhaus algorithm. It requires writing in memory four S_i, in other words $4 \times r \times d \times m_b$ bytes.

Furthermore, for these pre-computations, the private key's support F (d coefficients) and the support of the syndrome S ($r \times d$ coefficients) are needed.

Thus, the average memory cost of all these pre-computations is:

$$\text{Memory}_{pre-computed} = (r \times d \times (d+7) - 4 \times r + d) \times m_b.$$

With this formula, we can predict that ROLLO-I-128 requires 4,512 bytes to store the pre-computations which is too high for our chosen microcontroller. In order to reduce the memory cost, we store in memory at most three S_i and directly compute the two associated intersections as framed in Algorithm 5.

Algorithm 5: RSR (Rank Support Recover)

Input: $F = \langle f_1, \cdots, f_d \rangle$ an $\mathbb{F}_q$-subspace of $\mathbb{F}_{2^m}$, $s = (s_1, \cdots, s_n) \in \mathbb{F}_{2^m}^n$ the syndrome of an error e and r the rank's weight of e

Output: Vector subspace E

Compute $S = \langle s_1, \cdots, s_n \rangle_{\mathbb{F}_q}$

`// Recall that` $S_i = f_i^{-1}S$ `and` $S_{i,j} = S_i \cap S_j$

$tmp_1 \leftarrow S_1$

$tmp_2 \leftarrow S_2$

$tmp_3 \leftarrow S_3$

Compute $S_{1,2} = tmp_1 \cap tmp_2$

for i *from* 1 *to* $d-2$ **do**

  Compute $S_{i+1,i+2} = tmp_{i+1} \cap tmp_{i+2}$

  Compute $S_{i,i+2} = tmp_i \cap tmp_{i+2}$

  $tmp_{(i-1)\%3+1} \leftarrow S_{i+3}$

for i *from* 1 *to* $d-2$ **do**

  `// Direct sum of vector spaces`

  $tmp \leftarrow S + F \cdot (S_{i,i+1} + S_{i+1,i+2} + S_{i,i+2})$

  if $\dim(tmp) \leq rd$ **then**

    $S \leftarrow tmp$;

$E \leftarrow \bigcap_{1 \leq i \leq d} f_i^{-1} \cdot S$

return E

After the modifications, the total memory cost is:

$$\text{Memory}_{pre-computed} = (10 \times r \times d - 4 \times r + d) \times m_b.$$

This method allows us to save $(d-3) \times r \times d \times m_b$ bytes. The gains in memory for each security level are presented in the Table 4.

Table 4. Memory gains with the modified RSR algorithm

Algorithm	Bytes	Save bytes
ROLLO-I-128	3,432	1,080
ROLLO-I-192	4,836	2,016
ROLLO-I-256	8,100	4,480

2.3 Performance Evaluation

The cryptosystem is implemented in C. We target a microcontroller, based on a widely used 32-bit SecureCore® SC300™, which has an embedded 32-bit mathematical crypto co-processor to perform operations in $GF(p)$ and $GF(2^m)$ and a True Random Number Generator (TRNG). Among the 24 kB of RAM featuring on the microcontroller only 4 kB are available for cryptographic computations.

For performance measurements, we use IAR Embedded Workbench IDE for ARM [10] compiler C/C++ with high-speed optimization level.

We count the number of cycles with the debugging functionality of IAR.

An element in $\mathbb{F}_{2^m}^n$ is represented by $n \times \lceil m/32 \rceil \times 4$ bytes. For ROLLO-I-128, $m = 79$ and for ROLLO-I-192, $m = 89$, so we obtain $\lceil 79/32 \rceil = \lceil 89/32 \rceil = 3$ 32-bit words for both. Thus, the memory usages for ROLLO-I-128 and ROLLO-I-192 only differ according to n. Nevertheless, for ROLLO-I-256, $\lceil 113/32 \rceil = 4$, which explains the significant difference of memory usage between the higher security level and the two lower ones in the Table below.

In Table 5, the memory usage refers to the RAM required to perform the cryptosystem. The keys being stored in the EEPROM (Electrically Erasable Programmable Read-Only memory), we provide the memory usage with and without counting the public and secret keys. As we can see, ROLLO-I-256 cannot be implemented in our target device because its memory usage exceeds significantly the 4 kB of RAM but it can be executed with 8 kB of RAM, that is still reasonable.

Table 5. Memory usage for ROLLO-I (in bytes)

Security	Algo.					
	With keys			Without counting the keys		
	GenKey	Encap	Decap	GenKey	Encap	Decap
ROLLO-I-128	3,520	3,592	3,964	2,940	2,940	3,320
ROLLO-I-192	4,120	4,188	5,096	3,448	3,432	4,334
ROLLO-I-256	7,440	7,152	8,992	6,288	5,872	7,776

All the operations in $\mathbb{F}_2[x]/(P_m)$ take advantage of the crypto co-processor in $GF(2^m)$, leading the implementations using the crypto co-processor of ROLLO-I-128 and ROLLO-I-192 to be faster than their full software versions. We provide in Table 6 the number of cycles and the time in milliseconds required by ROLLO-I for the different security levels with the microcontroller running at 50 MHz.

In this paper, we do not compare our implementations with the reference implementation as it does not fit into the target microcontroller.

Table 6. Execution time of ROLLO-I

Security		Full software on SC300			On SC300 with co-processor		
		GenKey	Encap	Decap	GenKey	Encap	Decap
ROLLO-I-128	cycles ($\times 10^6$)	15.47	1.99	4.31	8.68	0.55	3.75
	ms	309	40.8	86.3	173.6	11	75
ROLLO-I-192	cycles ($\times 10^6$)	21.31	3.38	7.8	11.11	0.8	6.63
	ms	426	67.6	156	222.2	16	132.6
ROLLO-I-256	cycles ($\times 10^6$)	39.92	6.62	15.54	ND	ND	ND
	ms	798.5	132.5	310.8	ND	ND	ND

To see if ROLLO-I can be a realistic alternative to the current key exchange schemes, we compare in Table 7 the full software implementations with Elliptic Curve Diffie-Hellman key exchange (ECDH) [11] that is integrated in the same platform.

For ROLLO-I, the key agreement takes into account the Encapsulation and Decapsulation processes. As a remainder, for ECDH, two entities compute two scalars multiplication over $E(\mathbb{F}_q)$ in parallel to establish a shared secret. Thus, for its cost's estimation, we executed the scalar multiplications.

Table 7. Performance comparison between ROLLO-I and ECDH for two different security levels.

Security	Algorithm	Clock cycle ($\times 10^6$)
128	ROLLO-I-128	6.3
	ECDH Curve 256	3.49
192	ROLLO-I-192	11.18
	ECDH Curve 384	8.45

We observe that the two implementations are of the same order of magnitude.

3 Side-Channel Attack on ROLLO-I

Side-channel attacks were first introduced by Kocher in 1996 [12]. Some of these attacks exploit the leakage information coming from a device executing a cryptographic protocol. An adversary extracts these information without having to tamper with the device.

In this section, we deal with chosen-ciphertext Simple Power Analysis (SPA) attack. With the observation of the power traces, SPA attack consists of identifying sequences of an algorithm. This analysis leads a side-channel attacker to compute the secret key, used to establish the shared secret between two entities.

3.1 Attack

ROLLO-I does not require the use of ephemeral keys. The generation of keys is generally performed once in the life cycle of a component. The Encapsulation and Decapsulation processes are performed several times with the same key pair $((\mathbf{x}, \mathbf{y}, \mathbf{F}), \mathbf{h})$.

The decapsulation process is a good target for side-channel attacks because it involves the secret key $\mathbf{x}$ during the syndrome computation

$$\mathbf{s} = \mathbf{x} \cdot \mathbf{c} \bmod P_n.$$

Then, the aim of the attack is to recover the syndrome. The syndrome's support computation S applies Gaussian elimination algorithm to the matrix associated to the syndrome $\mathbf{s}$. The standard Gaussian elimination on a binary matrix is given in Algorithm 6.

Algorithm 6: Gaussian elimination algorithm

Input: Matrix $M \in \mathcal{M}_{n,m}(\mathbb{F}_2)$
Output: Matrix M under row echelon form and the rank of the matrix

```
1  Rank ← 0
2  for i = 0 to m − 1 do
3      for j = i to n − 1 do
4          if M_{j,i} = 1 then                      // Non-zero element - pivot
5
6              // The row j is a pivot
7              row i ↔ row j
8              Rank ← Rank +1
9              break
10     for k = row i + 1 to n do
11         if M_{k,i} = 1 then                      // row treatment
12
13             row k ← row k + row i
14 return (M, Rank)
```

The first non-zero coefficient in the column is the pivot. With the first for loop (line 3 – Algorithm 6) we scan each coefficient in the column to find the pivot.

Then we exchange the current row of the founded coefficient with the pivot row. The time required to determine the pivot indicates the number of coefficients processed and allows us to recover the pivot row.

With the second for loop (line 10 - Algorithm 6), we remove the other non-zero coefficients in the column.

Specifically, two different treatments are performed on each coefficients:

1. If the coefficient is 0 then no operation is performed.
2. If the coefficient is 1 then an addition in $\mathbb{F}_2[x]/(P_m)$ is performed between the pivot row and the one processed.

This difference of treatment leads us to determine the rows where are the non-zero coefficients. The syndrome's rank is at most $r \times d$. Thus, at the end of the process, we obtain a matrix M_s in row echelon form with the first column known by the attacker.

$$
M_s = \begin{pmatrix}
s_{0,0} & * & * & * & * & * \\
s_{1,0} & s_{1,1} & * & * & * & * \\
\vdots & \vdots & \ddots & * & * & * \\
s_{n-1,0} & s_{n-1,1} & \cdots & s_{n-1,r \times d-1} & * & *
\end{pmatrix}
$$

As the matrix is not under its reduced form and since it's a binary matrix, the systems of equations should be complicated to solve. That is why we only consider the first column for the attack.

To recover the syndrome, we perform m rotations of the matrix M_s with the use of the initial ciphertext. Namely, we multiply the ciphertext $\mathbf{c}$ by x^i in $\mathbb{F}_{2^m}[X]/(P_n)$, with $1 \leq i < m$.

However, we have to consider the modular rotation during the recovering of the columns' syndrome matrix. Considering ROLLO-I-128 parameters given in Table 2. Multiplying the ciphertext by x in $\mathbb{F}_{2^{79}}[X]/(P_{47})$ implies that the last column of the matrix syndrome is xored with the columns 0 and 9 as depicted in Fig. 3. This is due to the modulo $P_{79}(x) = x^{79} + x^9 + 1$ involved in the field $\mathbb{F}_{2^{79}}[X]/(P_{47})$.

$$
\begin{pmatrix}
0 & s_{0,0} & s_{0,1} & \cdots & s_{0,8} & \cdots & s_{0,77} \\
0 & s_{1,0} & s_{1,1} & \cdots & s_{1,8} & \cdots & s_{1,77} \\
\vdots & \vdots & \vdots & \cdots & \vdots & \cdots & \vdots \\
0 & s_{46,0} & s_{46,1} & \cdots & s_{46,8} & \cdots & s_{46,77}
\end{pmatrix}
\begin{matrix}
s_{0,78} \\
s_{1,78} \\
\vdots \\
s_{46,78}
\end{matrix}
$$

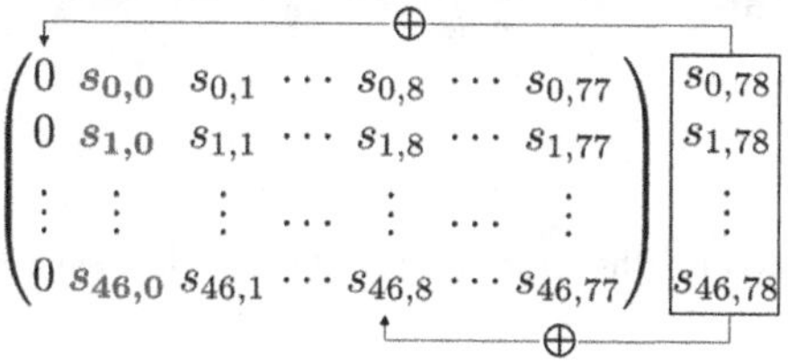

Fig. 3. Example of modular rotation for the syndrome's matrix for ROLLO-I-128

The column 78 is recovered as explained above and to recover the column 9 xored with column 78, we multiply the ciphertext by x^{69} modulo $P_{79}(x)$.

In ROLLO-I-128 and ROLLO-I-256, we need to keep in mind the xor when recovering the columns 1 to 8. For ROLLO-I-192, columns 1 to 38 are concerned by a xor.

To develop this attack, we target the implementation using the co-processor. For the experiment, we consider the parameters of ROLLO-I-128, namely n = 47, and m = 79. The secret key **x** and the ciphertext **c** involved in the syndrome computation are generated during the Key Generation and Encapsulation processes. ROLLO-I-128 traces are captured with a Lecroy SDA 725Zi-A oscilloscope. We observe in Fig. 4 the difference of patterns between the treatment of bits 1 and 0. This trace allows us to recover the first column of the syndrome's matrix corresponding to

$$10110101110111010001010111001111001001110010110.$$

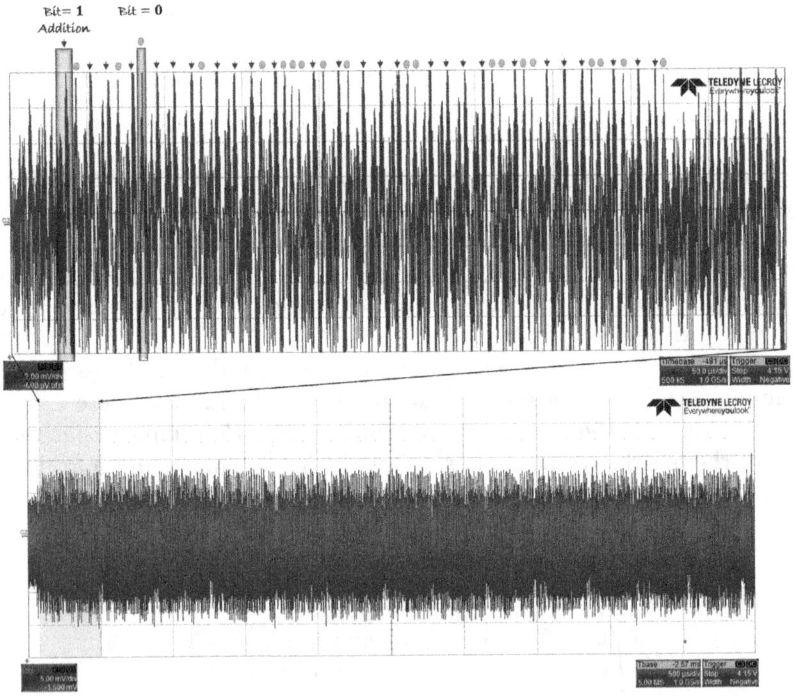

Fig. 4. SPA performed on the first column during Gaussian elimination process

We use the same technique to recover all the columns after the matrix rotation and finally the syndrome.

3.2 Countermeasures

Let us discuss solutions to secure the cryptosystem against the attack explained previously. Several solutions are available to protect the Gaussian elimination against SPA attacks.

Two solutions consist of:

- implementing a constant-time algorithm in which the additions in $\mathbb{F}_2[x]/(P_m)$ are independent from the processed coefficients, as presented in [13]. Thus, the execution time is no longer depending on the private key;
- adding dummy operations when processing the coefficients 0 as described in Algorithm 7.

These solutions require an additional element in $\mathbb{F}_{2^m}$ to store intermediate results.

Algorithm 7: Gaussian elimination algorithm with dummy operations

Input: Matrix $M \in \mathcal{M}_{n,m}(\mathbb{F}_2)$

Output: Matrix M in row echelon form and the rank of the matrix

```
Rank ← 0
Temp ← 0
for i = 0 to m − 1 do
    for j = i to n − 1 do
        if M_{j,i} = 1 then
            // The row j is a pivot
            row i ↔ row j
            Rank ← Rank +1
            break
    for k = row i + 1 to n do
        if M_{k,i} = 1 then
            row k ← row k + row i
        else
            Temp ← row k + row i
return (M, Rank)
```

The Fig. 5 presents the trace of Gaussian elimination with dummy operations. We observe a uniformization of the trace due to the added noises.

With the proposed countermeasures, an attacker is not able to exploit patterns according to the processed bit. However, these solutions are subjected to other side-channel attacks that are not covered in this paper such as Differential Power Analysis (DPA) attacks.

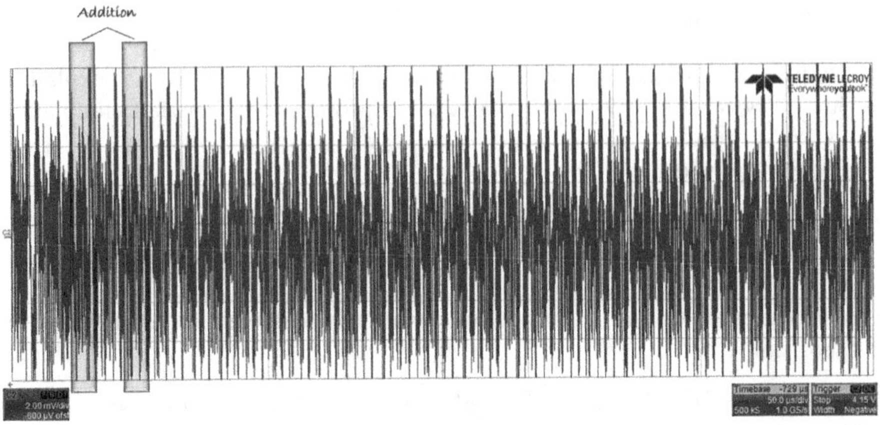

Fig. 5. Trace for the first column in Gaussian elimination with dummy operations.

Another solution is to randomize the treatment of coefficients in each column as described in Algorithm 8. An attacker is not able to recover the indices of pivots and processed rows. Considering the first column, the attacker has n possibilities for the pivot and $(n-1)!$ possibilities for the row treatment. Thus, the complexity of the SPA attack is $(n!)^m$. For example, with ROLLO-I-128, the complexity is $(47!)^{79}$ which corresponds to about $2^{15,591}$ possibilities.

Algorithm 8: Gaussian elimination with randomization

Input: Matrix $M \in \mathcal{M}_{n,m}(\mathbb{F}_2)$
Output: Matrix under row echelon form and the rank of the matrix

Rank $\leftarrow 0$
for $i = 0$ *to* $m-1$ **do**
    for $j = i$ *to* $n-1$ **do**
        $j_{rand} = (j + random()) \mod (n-i)$ `// randomization`
        if $M_{j_{rand},i} = 1$ **then**
            `// The row` j_{rand} `is a pivot`
            row $i \leftrightarrow$ row j_{rand}
            Rank $\leftarrow$ Rank $+1$
            break
    for $k = row\ i+1$ *to* n **do**
        $k_{rand} = (k + random()) \mod (n-k)$ `// randomization`
        row $k \leftrightarrow$ line k_{rand}
        if $M_{k,i} = 1$ **then**
            row $k_{rand} \leftarrow$ row k_{rand} + row i
return $(M, Rank)$

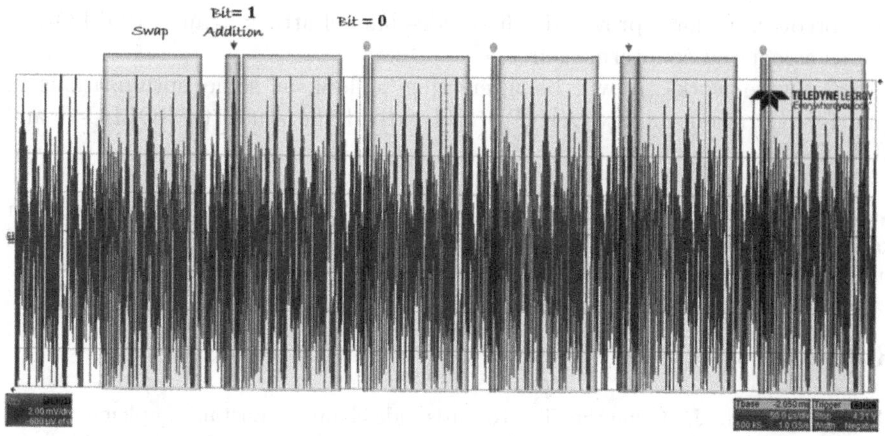

Fig. 6. Trace of the first column in Gaussian elimination process after application of randomization

Figure 6 provides the trace of the execution of Gaussian elimination with the countermeasure presented in Algorithm 8.

Although we can still distinguish the coefficients 0 and 1, the order of elements in each column is completely random so we can no longer exploit this information.

As we can see in Table 8, regarding the randomization in Gaussian elimination process, exchanging two rows at each iteration has a significant impact on the execution time of decapsulation process, increasing it by about 50%. The second countermeasure impacts the execution time by about 40%.

Table 8. Executing time of ROLLO-I decapsulation with countermeasures.

Security		Decapsulation		
		With randomization	With dummy operation	Without countermeasures
128	cycles ($\times 10^6$)	8.09	5.84	4.31
	ms	161.8	116.6	86.3
192	cycles ($\times 10^6$)	17.01	11.23	7.8
	ms	340.2	224.6	156
256	cycles ($\times 10^6$)	32.45	21.62	15.54
	ms	649	432.4	310.8

Conclusion

In this paper, we have highlighted that ROLLO-I can be implemented in a chosen constraint device and the structure used allows the cryptosystem to benefit from the current crypto co-processor. We have also shown that in comparison with existing algorithms such as ECDH, our implementation's performances were compelling.

Moreover, we have provided a first side-channel attack against ROLLO-I as well as associated countermeasures.

For future works, it will be interesting to look at some optimizations in time for operations in $\mathbb{F}_{q^m}[X]/(P_n)$ and extend the study to ROLLO-II and ROLLO-III.

Acknowledgements. We would like to thank the CBCrypto 2020 reviewers for their valuable comments and suggestions. We also thank Steve Clark and Tania Richmond for proofreading the revised version of the paper and for useful discussions.

References

1. von Maurich, I., Güneysu, T.: Towards side-channel resistant implementations of QC-MDPC McEliece encryption on constrained devices. In: Mosca, M. (ed.) PQCrypto 2014. LNCS, vol. 8772, pp. 266–282. Springer, Cham (2014). https://doi.org/10.1007/978-3-319-11659-4_16
2. Richmond, T., et al.: A side-channel attack against the secret permutation on an embedded McEliece cryptosystem. In: 3rd Workshop on Trustworthy Manufacturing and Utilization of Secure Devices - TRUDEVICE 2015, Grenoble, France (2015). https://hal-ujm.archives-ouvertes.fr/ujm-01186639
3. Gaborit, P., et al.: Low rank parity check codes and their application to cryptography, April 2013
4. Aguilar Melchor, C., et al.: NIST PQC second round submisssion: ROLLO - Rank-Ouroboros, LAKE & LOCKER (2019)
5. National Institute of Standards and Technology. Submission Requirements and Evaluation Criteria for the Post-Quantum Cryptography Standardization Process (2016). https://csrc.nist.gov/CSRC/media/Projects/Post-Quantum-Cryptography/documents/call-for-proposals-final-dec-2016.pdf
6. Hankerson, D., et al.: Guide to Elliptic Curve Cryptography. Springer, Heidelberg (2003)
7. Cohen, H., et al.: Handbook of Elliptic and Hyperelliptic Curve Cryptography. Discrete Mathematics and Its Applications. CRC Press, Boca Raton (2005)
8. Weimerskirch, A., Paar, C.: Generalizations of the Karatsuba Algorithm for Efficient Implementations (2006). aweimerskirch@escrypt.com 13331. Accessed 2 July 2006. http://eprint.iacr.org/2006/224
9. Luks, E.M., et al.: Some algorithms for nilpotent permutation groups. J. Symb. Comput. **23**(4), 335–354 (1997). https://doi.org/10.1006/jsco.1996.0092
10. IAR Embedded Workbench. https://www.iar.com/
11. SEC 1. Standards for Efficient Cryptography Group: Elliptic Curve Cryptography - version 2.0 (2009). https://www.secg.org/sec1-v2.pdf
12. Kocher, P.C.: Timing attacks on implementations of Diffie-Hellman, RSA, DSS, and other systems. In: Koblitz, N. (ed.) CRYPTO 1996. LNCS, vol. 1109, pp. 104–113. Springer, Heidelberg (1996). https://doi.org/10.1007/3-540-68697-5_9
13. Aguilar Melchor, C., et al.: Constant-time algorithms for ROLLO (2019). https://csrc.nist.gov/CSRC/media/Events/Second-PQC-Standardization-Conference/documents/accepted-papers/caullery-constant-time-rollo.pdf
14. Aragon, N., et al.: NIST PQC second round submisssion: BIKE - Bit Flipping Key Encapsulation (2019)

15. Aguilar Melchor, C., et al.: NIST PQC second round submisssion: Hamming Quasi-Cyclic (HQC) (2019)
16. Aguilar Melchor, C., et al.: NIST PQC second round submisssion: Rank Quasi-Cyclic (RQC) (2019)
17. McEliece, R.J.: A public-key cryptosystem based on algebraic coding theory. Deep Space Netw. Prog. Rep. **44**, 114–116 (1978)

Author Index

Printed in the United States
By Bookmasters